T0310649

Design and Development of Smart Surgical Assistant Technologies

Are Amazon Alexa and Google Home limited to our bedrooms, or can they be used in hospitals? Do you envision a future where physicians work hand-in-hand with voice AI to revolutionize healthcare delivery? In the near future, clinical smart assistants will be able to automate many manual hospital tasks – and this will be only the beginning of the changes to come.

Voice AI is the future of physician–machine interaction, and this focus book provides invaluable insight into its next frontier. It begins with a brief history and current implementations of voice-activated assistants and illustrates why clinical voice AI is at its inflection point. Next, it describes how the authors built the world's first smart surgical assistant using an off-the-shelf smart home device, outlining the implementation process in the operating room. From quantitative metrics to surgeons' feedback, the authors discuss the feasibility of this technology in the surgical setting. The book then provides an in-depth development guideline for engineers and clinicians desiring to develop their own smart surgical assistants. Lastly, the authors delve into their experiences in translating voice AI into the clinical setting and reflect on the challenges and merits of this pursuit.

The world's first smart surgical assistant has not only reduced surgical time but eliminated major touch points in the operating room, resulting in positive, significant implications for patient outcomes and surgery costs. From clinicians eager for insight into the next digital health revolution to developers interested in building the next clinical voice AI, this book offers a comprehensive guide for both audiences.

Design and Development of Smart Surgical Assistant Technologies

A Case Study for Translational Sciences

Jeff J. H. Kim, Richard Um, Rajiv Iyer,
Nicholas Theodore, and Amir Manbachi

CRC Press
Taylor & Francis Group
Boca Raton London New York

CRC Press is an imprint of the
Taylor & Francis Group, an **informa** business

First edition published 2023
by CRC Press
6000 Broken Sound Parkway NW, Suite 300, Boca Raton, FL 33487-2742

and by CRC Press
4 Park Square, Milton Park, Abingdon, Oxon, OX14 4RN

CRC Press is an imprint of Taylor & Francis Group, LLC

ISBN: 9781032168722 (hbk)
ISBN: 9781032181967 (pbk)
ISBN: 9781003253341 (ebk)

DOI: 10.1201/9781003253341

Typeset in Times
by Deanta Global Publishing Services, Chennai, India

We dedicate this book to our families and our loved ones.

CONTENTS

ACKNOWLEDGMENTS

The authors would like to acknowledge our invaluable team members, Jonathan Liu and Japesh Patel, for their part in engineering the smart surgical assistant and the faculty from the Johns Hopkins School of Medicine Department of Neurosurgery, Department of Radiology, and Department of Biomedical Engineering. We would like to also thank Gina Lynn Adrales, M.D., M.P.H., Ivan George, and Nick Louloudis from MISTIC at Johns Hopkins University. We acknowledge Clare Sonntag for her edits. Last, but not least, we thank Carolina Antunes and Betsy Byers at CRC Press for this amazing opportunity and for bringing this book to reality.

Nicholas Theodore and Amir Manbachi acknowledge funding support from Defense Advanced Research Projects Agency, DARPA, Award Contract #: N660012024075. In addition, Amir Manbachi acknowledges funding support from Johns Hopkins Institute for Clinical and Translational Research (ICTR)'s Clinical Research Scholars Program (KL2), administered by the National Center for Advancing Translational Sciences (NCATS), National Institute of Health (NIH).

AUTHOR BIOGRAPHIES

Jeff J.H. Kim is an MD/PhD student at the University of Illinois College of Medicine, Chicago, Illinois, leading developments in medical AI, health technology, and organ-on-chip microfluidics. During his master's program at Johns Hopkins University, Baltimore, Maryland, he spearheaded the development of a voice-activated surgical assistant, where he successfully implemented his pioneering work in the operating room, improving both patient safety and surgical efficiency. His work is published in the 2021 SPIE Medical Imaging Conference Proceedings. He also led the development of several innovative works in neurosurgery and cardiology, developing a voice-activated smart surgical bed and cardiac patch that detects lethal arrhythmia. He obtained a dual bachelor's degree in Electrical Engineering and Neuroscience at Johns Hopkins University.

Richard Um is a graduate researcher at the Johns Hopkins University School of Medicine. His research focuses on translational neurosurgery with an emphasis on in-vitro benchtop models to test certain devices used in surgical procedures. With his extensive design and development skills, he created the first prototype of a working voice-controlled operating bed. He has further developed a smart hospital assistant that provides surgeons with complete control over equipment in the operating room, prioritizing patient safety and workflow efficiency. Richard Um received a bachelor's degree in Biomedical Engineering with a minor in Robotics from Johns Hopkins University, as well as a master's in Biomedical Sciences from Tufts University, Medford, Massachusetts.

Rajiv Iyer is a pediatric neurosurgeon who recently completed a neurosurgical residency at The Johns Hopkins Hospital and a pediatric neurosurgery fellowship at Primary Children's Hospital/University of Utah, Salt Lake City, Utah, and is currently completing an Advanced Pediatric Spinal Deformity Fellowship at Columbia University/ Morgan Stanley Children's Hospital, New York City, New York, under the mentorship of Dr. Lawrence Lenke and Dr. Michael Vitale. Dr. Iyer will soon be joining the pediatric neurosurgery group at the University of Utah as Assistant Professor of Neurosurgery. There, his clinical focus will be on the treatment of complex pediatric spinal disorders, including spinal deformity, craniocervical junction disorders, and spinal column/spinal cord tumors. Thus far in his career, Dr. Iyer has been passionate about learning from experts in the field and utilizing the best possible techniques to care for his patients. He is enthusiastic to begin his academic neurosurgical career, where he hopes to deliver outstanding pediatric neurosurgical care, while adopting new technology in and out of the operating room in an effort to improve patient outcomes and advance the field.

Nicholas Theodore is an American neurosurgeon and researcher at Johns Hopkins University School of Medicine. He is known for his work in spinal trauma, minimally invasive surgery, robotics, and personalized medicine. He is Director of the Neurosurgical Spine Program at Johns Hopkins and Co-Director of the Carnegie Center for Surgical Innovation and Co-Director of the HEPIUS Neurosurgical Innovation Lab at Johns Hopkins.

In 2016 he became the second Donlin M. Long Professor of Neurosurgery at Johns Hopkins Hospital. Dr. Theodore also holds professorships in Orthopedics and Biomedical Engineering at Johns Hopkins. He is actively involved in the area of preventative medicine within neurosurgery. He has been associated with the ThinkFirst Foundation for several years, having served as the foundation's Medical Director and President. In 2017, Dr. Theodore was appointed to the National Football League's Head, Neck and Spine Committee, of which he became Chairman in 2018. In 2020, Michael J. Fox revealed in his memoir that Dr. Theodore performed a risky but successful surgery on him to remove an ependymoma in Fox's spinal cord.

In 2020, Dr. Theodore received a grant in the amount of $13.48 million from the Defense Advanced Research Projects Agency's (DARPA) Bridging the Gap Plus (BG+) program to fund research in new approaches to the treatment of spinal cord injury. With this grant, Dr. Theodore is co-leading the effort to treat patients with spinal cord injury by integrating injury stabilization, regenerative therapy, and functional restoration using targeted electrical ultrasound modalities. As the principal investigators for the program, Dr. Theodore and Dr. Manbachi will coordinate teams at Johns Hopkins and its Applied Physics Laboratory, Columbia University, and Sonic Concepts.

Dr. Amir Manbachi is an Assistant Professor of Neurosurgery, Biomedical Engineering, Mechanical Engineering, Electrical and Computer Engineering at Johns Hopkins University. He is the Co-Founder and Co-Director of the HEPIUS Neurosurgical Innovation Lab. His research interests include applications of sound and ultrasound to various neurosurgical procedures. These applications include imaging the spine and brain, detection of foreign body objects, remote ablation of brain tumors, monitoring of blood flow and tissue perfusion, and other upcoming interesting applications such as neuromodulation and drug delivery. His pedagogical activities have included teaching engineering design, innovation, translation, and entrepreneurship, as well as close collaboration with clinical experts in Surgery and Radiology at Johns Hopkins.

Dr. Manbachi is an author of over 25 peer-reviewed journal articles, over 30 conference proceedings, over 10 invention disclosures/patent applications, and a book entitled *Towards Ultrasound-Guided Spinal Fusion Surgery*. He has mentored more than 150 students and has so far raised $15 million in funding, and his interdisciplinary research has been recognized by a number of awards, including the University of Toronto's 2015 Inventor of Year award, the Ontario Brain Institute 2013 fellowship, the Maryland Innovation Initiative, and the Johns Hopkins Institute for Clinical and Translational Research's Career Development Award.

Dr. Manbachi has extensive teaching experience, particularly in the fields of engineering design, medical imaging, and entrepreneurship (at both Hopkins and Toronto), for which he has received numerous awards: the University of Toronto's Teaching Excellence award (2014), the Johns Hopkins University career center's award nomination for students' "Career Champion" (2018), and the Johns Hopkins University Whiting School of Engineering's Robert B. Pond Sr. Excellence in Teaching Excellence Award (2018).

INTRODUCTION TO VOICE-ACTIVATED ASSISTANTS

When people think of artificial intelligence (AI), many envision virtual butlers who are capable of handling anything we ask of them. We can attribute this impression to the mass media —consider Jarvis from Marvel's Iron Man series. From manufacturing Iron Man's suits to helping him fight his enemies, Jarvis does it all under Tony Stark's orders. Although voice-activated smart assistants fail to encompass the vastness that is AI, they have undoubtedly played a vital role in familiarizing people with it.

Of all the AI applications available in the world today, voice-assistant technology has been one of the most pervasive and widespread. Today, over half of Americans interact with virtual assistants embedded in their smartphones and many have one or more consumer smart home devices.[1] However, this was not the norm even a decade ago. Hardware and software advancements have allowed for rapid growth and expansion of voice-activated assistant technology. Once only depicted in science-fiction novels, voice-activated assistants have evolved into a common household technology. What contributed to this rise? Where will the technology go from here?

1.1 WHAT IS VOICE-ACTIVATED TECHNOLOGY?

Before we move forward, we must set the operational definition of *voice-activated technology*. Here, voice-activated technology is any technology capable of executing pre-programmed tasks based on vocal input by a user. This technology goes a step beyond speech-recognition technology, as it can not only understand users' requests but can also deliver convenience by executing user-specified commands.

To be considered a voice-activated smart assistant, the technology must satisfy three main criteria. First, it must be able to capture and decode human speech. This is the human equivalent of comprehension. Second, the technology needs to carry out a plurality of tasks that offer convenience to the user. This is what makes them "smart assistants". Third, it requires a human–machine interface, which often takes the form of a smart device that facilitates the interaction between the AI and the user.

1.2 SYSTEM ARCHITECTURE OF VOICE-ACTIVATED TECHNOLOGY

The proliferation of smart voice-activated assistant technology would not have been possible without the foundational bedrock that supports its existence. The most important support frameworks are the internet of things (IoT) and Natural Language Processing (NLP) technology. IoT gives the physical assembly and the network access for smart voice-activated assistants. It also allows voice-activated technology to connect to other IoT devices, allowing functional flexibility. In 2018, the number of connected IoT devices reached 22 billion around the world and that number is still growing.[2] And this feat is made possible by the maturing technology in the space of wireless connectivity, battery, integrated circuit, and cloud computing. We will briefly dive into each of these components (Figure 1.1).

First and foremost, wireless technology, specifically, Wi-Fi, low-energy Bluetooth, and Low Power Wide Area Network (LPWAN), established a communication protocol that connects machine to machine (M2M), allowing for innovative and hassle-free device interconnectivity.[3] Second, the development and expansion of rechargeable lithium-ion batteries revolutionized electronics development as it allowed for compact and portable electronics like wearables and IoT devices. Next, the invention of the integrated circuit (IC) gave rise to the compact housing of transistors leading to advanced microprocessors. The role of IC is reviewed more extensively in Section 1.3.3.1. Lastly, cloud computing granted IoT devices to take advantage of remote computer system resources like computing power and data, reducing the on-board hardware requirements. Each of these elements contributed to a flexible, mobile, and compact electronic arrangement, which is deemed critical to the success of IoT devices.

The other crucial constituent of a smart voice-activated assistant is NLP. Today's NLP effort is made possible by the advancements in speech recognition and machine learning beginning in the 20th century. For the modern voice-activated assistants, however, it is

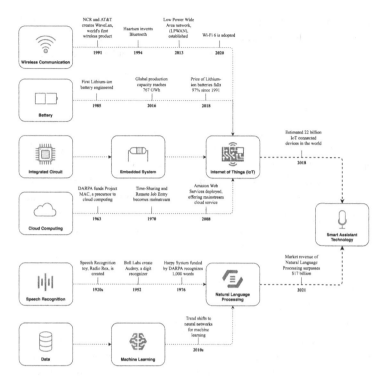

Figure 1.1 A flowchart that illustrates the components and their respective key development milestones that allowed for the birth of voice-activated assistant technology. Simply put, the voice-activated assistant technology is a convergence of internet of things (IoT) technology that gives it a flexible framework and Natural Language Processing (NLP) that allows it to comprehend and speak human language. IoT is comprised of four different components – wireless technology, battery, cloud computing, and integrated circuit. NLP is a combination of speech-recognition technology and machine learning.

the transition from the statistical NLP to the neural NLP, a neural network-dependent machine learning, that gave the greatest contribution.[4] The efforts in speech recognition are explored in greater detail below.

The voice-activated assistant technology would not have been made possible if any one of the supporting frameworks was absent. And because these supporting frameworks have reached their maturity, the development of a diverse smart voice-activated assistant application is the most favorable today. It is important to keep the system architecture in mind as we move forward and understand how these building blocks interact with one another. It would be useful when we dive into the engineering process of the smart surgical assistant.

1.3 HISTORY

Because major advancements in commercial voice-activated smart assistant technology occurred just in the past decade, it is easy to mislabel the technology as a recent development. However, this could not be farther from the truth. The effort to create voice-activated technology dates back as early as the 1920s. It is important to understand how voice-activated assistants have evolved in order to anticipate the future trajectory of their development. Looking at what has been done can give us an idea of where the technology can go from here.

1.3.1 Humble Beginnings: The 1920s

1.3.1.1 Radio Rex

The very first voice-activated technology was neither smart nor particularly useful, but it did spark joy among the masses in the early 1920s. Twenty years before the advent of the first computer, Elmwood Button Co. produced a toy called Radio Rex – a toy dog that crawled out of its home when its name, "Rex", was called. The mechanism of this toy, though simple, was quite clever. The acoustic energy in the word "Rex", specifically the vowel [eh], triggered a harmonic oscillator that released Rex from a current-energized magnet. The frequency detector in Radio Rex, however, had its shortcomings: it would respond to other words at 500 Hz frequency. It also had trouble detecting the vocal frequency of children and females. However, this quirky toy would mark the beginning of using vocal frequency as part of speech recognition that would guide future developments[5] (Figure 1.2).

1.3.2 Significant Developments in Speech Recognition: 1950 to 1990

1.3.2.1 Digit Recognizers

Fast forward 30 years to 1952. Bell Labs introduced Audrey, a digit recognizer that stood 6 feet tall and contained analog filters and circuitry. Despite its enormous size, Audrey boasted the ability to

Figure 1.2 Photo of Radio Rex, the earliest known voice-activated electronics. A dog toy crawls out of its house when its name "Rex" is called out by the user. The specific frequency in the word "Rex" triggers a harmonic oscillator, which releases the dog figure from the current-energized magnet. Image used with permission of The Warden and Scholars of Winchester College.

recognize just ten numbers, from 0 to 9. The operator would speak into a built-in telephone, and Audrey would match the speech sounds to pre-referenced electrical buses stored within an analog memory. Flashing light indicated a visible output. The system had its short-comings; for example, the analog reference memory had to be tailored to the operator, restricting the number of users. However, once paired with an operator, Audrey boasted a 97% accuracy rate. Its ability to recognize ten digits was enough to showcase the untapped potential of speech recognition technology. Scientists, engineers, and the general public alike were fascinated by a non-living entity processing the complexity that is human speech. Audrey's greatest legacy is the scores of developments that took place to expand speech-recognition technology, which lay the foundation for voice-activated smart assistants.[6,7]

1.3.2.2 DARPA

One key development that followed Audrey is the Speech Understanding Research (SUR) Program. Launched in 1976, the SUR program was funded by the Defense Advanced Research Projects Agency (DARPA) – $15 million was dedicated to building a system that could understand 1,000 words, or that would be equivalent to the vocabulary of a three-year-old. The initial design goal was satisfied by the Harpy System, which was developed at Carnegie Mellon. Harpy understood over 90% of spoken sentences from a predetermined 1,000-word database. This was a significant jump from Audrey, and Harpy showed improvements in the spectrum and number of words recognized. From this DARPA program, scientists and engineers introduced a guideline to develop the next generation of speech-recognition systems. They envisioned scaling existing techniques, such as linear predictive coding, dynamic time warping, and hidden Markov models. The projects that followed the SUR program also expanded on applying neural networks for automatic speech recognition.[3] We will not dive into these techniques, as such an analysis would deviate from the focus of this book. However, readers can refer to the further readings listed below if they are interested in learning more.

1.3.3 Modern Developments: 1990 to 2020

1.3.3.1 Development of Internet, Microprocessors, and Internet of Things Devices

The significant expansion of the internet and advanced microprocessor technology has enabled the rise of next-generation voice-activated assistants. Advancement in the microprocessor space enabled an exponential increase in processing power due to the rapid rise of transistor count.[8] As mentioned previously, the root of this phenomenon was the rise of system-on-chip circuits, also known as IC. Instead of utilizing motherboard-based PC architecture, which separates components based on functionality, integrated circuits allowed the consolidation of main and peripheral processing cores in a compact form. This had two main benefits that ultimately laid the groundwork for IoT technology.[9] The first is the significant reduction of size and vast improvement in computing and battery performance. Second, the rise of the modern internet and short-range wireless technology demonstrated the value of IoT technology, as access to the internet and device interconnectivity expanded the scope of voice-activated smart assistant capabilities. Short-range wireless technology such as Bluetooth advanced inter-device communication, while faster data transfer protocol expanded the scope of control over sensors and actuators.[10] This, in simple terms, gave arms and legs to the brain of the system, thereby facilitating smart functions such as controlling a

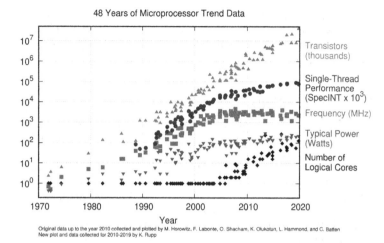

Figure 1.3 Trend data of microprocessor technology from 1970 to 2020. Transistor count has seen an exponential increase, which allowed for the advent of the internet of things technology. Single-thread performance, clock speed, and battery performance also increased rapidly until they tapered off starting in 2010. The core count has begun to see an increase starting in the mid-2000s, to compensate for the tapering of single-core performance and to allow for parallel processing. Source: https://github.com/karlrupp/microprocessor-trend-data, under a Creative Commons Attribution 4.0 International Public License.

thermostat with a smart home device. These two developments naturally gave rise to the highly capable and compact voice-activated smart assistant devices we know and love today (Figure 1.3).

1.3.3.2 Modern Voice-Activated Smart Assistant Technology

The greatest advancements in voice-activated smart assistant technology happened in the past decade. IBM Watson shows the pinnacle of this technology, as it stands as the most advanced voice-activated technology today. It can answer questions posed in natural language, boasting advanced automated reasoning. It became famous for winning the 2011 *Jeopardy!* match against champions Brad Rutter and Ken Jennings.[11] Consumer smart assistants have also made tremendous strides since their introduction in the 2010s, and have become essential tools for users in just a matter of a few years. Major tech companies such as Google and Amazon have invested heavily into developing their own voice-activated assistants in the hopes of connecting their services to more users. Voice-activated technology is only expected to grow bigger and more advanced. Voice will be the new, dominant human–computer interface, while typing becomes a relic of the past.

FURTHER READING

Gold, Ben, Nelson Morgan, and Dan Ellis. *Speech and audio signal processing: Processing and perception of speech and music*. John Wiley & Sons, 2011.

O'Shaughnessy, Douglas. "Linear predictive coding." *IEEE Potentials* 7.1 (1988): 29–32.

Müller, Meinard. "Dynamic time warping." *Information Retrieval for Music and Motion* (2007): 69–84.

Varga, A. P., and Roger K. Moore. "Hidden Markov model decomposition of speech and noise." *International Conference on Acoustics, Speech, and Signal Processing*. IEEE, 1990.

REFERENCES

1. Kinsella, Bret. "Smart home ownership nearing 50% of U.S. adults with voice assistant control becoming more popular – New research." *Voicebot.ai*, 18 December 2020, https://voicebot.ai /2020/12/18/smart-home-ownership-nearing-50-of-u-s-adults-with -voice-assistant-control-becoming-more-popular-new-reserach/.

2. Vailshery, Lionel Sujay. "Internet of Things (IoT) – Statistics & facts." *Statista*.

3. Chen, Min, Jiafu Wan, and Fang Li. "Machine-to-machine communications: Architectures, standards and applications." *KSII Transactions on Internet and Information Systems (TIIS)* 6.2 (2012): 480–497.

4. Goldberg, Yoav. "A primer on neural network models for natural language processing." *Journal of Artificial Intelligence Research* 57 (2016): 345–420.

5. David, E. E., and O. G. Selfridge. "Eyes and ears for computers." *Proceedings of the IRE* 50.5 (1962): 1093–1101.

6. Gold, Ben, Nelson Morgan, and Dan Ellis. *Speech and audio signal processing: Processing and perception of speech and music*. John Wiley & Sons, 2011.

7. Moskvitch, Katia. "The machines that learned to listen." *BBC Future* 15 (2017).

8. Schaller, Robert R. "Moore's law: Past, present and future." *IEEE Spectrum* 34.6 (1997): 52–59.

9. Alioto, Massimo, ed. *Enabling the Internet of Things: From integrated circuits to integrated systems*. Springer, 2017.
10. Kocakulak, Mustafa, and Ismail Butun. "An overview of wireless sensor networks towards internet of things." *2017 IEEE 7th Annual Computing and Communication Workshop and Conference (CCWC)*. IEEE, 2017.
11. Memeti, Suejb, and Sabri Pllana. "PAPA: A parallel programming assistant powered by IBM Watson cognitive computing technology." *Journal of Computational Science* 26 (2018): 275–284.

COMMERCIAL VOICE-ACTIVATED ASSISTANTS

2.1 REASON FOR ADOPTION

Analyzing the factors that led to the mainstream adoption of voice-activated technology can help us understand what may push healthcare professionals to adopt this technology in their practices. In this section, we will review the major advantages of voice-activated technology and analyze whether it has a place in the clinical setting.

2.1.1 Hands-Free Interaction

According to the Pew Research Center, 55% of consumers cited hands-free operation as the major reason for adopting voice-activated assistants. Having their hands free allows people to multitask, sparing them significant time and focusing power. Instead of manually using one's mobile device to search for the weather, one can simply ask a smart assistant to do the research. One important benefit of this human–machine interface is that the user does not have to withdraw themselves from the current task with which they are engaged. This can be useful in situations that require the user's full and undivided attention, such as driving. Indeed, the most commonly reported use for smartphone voice assistants is found while the user is driving, sitting at 62% of all use cases. With a person's hands on the steering wheel, they can use voice-activated assistants to serve as another set of arms.[1]

The use case is not limited to and during potentially hazardous activities. Voice-activated smart assistants are actively utilized in everyday situations. People report using this technology in hands-on activities like cooking or cleaning, as well as when their hands are unoccupied –while watching TV, for instance. The convenience of verbalizing commands to a smart assistant compared to the physical labor exerted by typing on a smartphone pushed many users to adopt this technology.

2.1.2 Voice Is the Most Natural Mode of Communication

Another common reason for the adoption of voice-activated smart assistants lies in the nature of vocal communication. Many users feel that spoken language is more natural than typing. This is very true if we consider how we learn and practice communicating. When we are born, the very first action we take – apart from kicking – is crying. We use our vocal cords to announce ourselves to the world, and our ability to deliver messages verbally is a vital tool and a survival mechanism we have adapted to use. Voice is the main communication modality most people utilize on a daily basis, and for good reason: our early ancestors learned to communicate by talking before any writing system was developed. This stands true today, as we still learn to talk before we learn to write. Voice is also special as it can be easily used across all age groups –children and elderly persons – alike. The same cannot be said about typing, which takes dedicated time and practice. Many elderly individuals fail to learn to type at all. The success in capturing the elderly population hints at a potential geriatric medical application of this technology, as we will discuss in later chapters.

2.1.3 Voice Is the Most Efficient Mode of Communication

Oftentimes, we find ourselves browsing through a long email chain that spans multiple days or even weeks and thinking, "This could have been a simple phone call". More often than not, it takes longer to deliver a message via email and/or text than via direct vocal communication. Conversing with others, whether it be on the phone or face to face, has zero to little lag time. We receive direct responses from the other party, which makes follow-up easy. Conversation allows us to address multiple topics and issues in one sitting, in a short span of time. In text, this is not the case – the flow of messages is often disrupted, leading to delays in conversations. Delays then lead to misunderstandings. When voice-activated assistants are used, communication is faster, which can allow for speedier processing and command execution. How can we use this to our advantage? Healthcare clinics can rely on voice-activated assistants for quicker processing of computer-based activities, such as drafting patient records. The average person can type 50 words per minute, whereas speech recognition software can easily transcribe more than three times that number.[2] Companies like Nuance and Suki are already establishing virtual health assistants that can automatically transcribe patient interactions and draft patient history.

Another main benefit of voice is that the volume, pitch, and intonation of our voice can give contextual cues in a conversation, which are crucial in understanding intent. Developers are currently going the extra mile to improve emotional intelligence in voice-activated assistants, allowing for more personalized and connected experience. Companies like OTO are trying to analyze users' emotions and energy levels, both of which are used to identify user satisfaction and generate emotionally accurate responses. As the potential uses of AI grow, we envision a future in which voice-activated assistants can have emotionally intelligent conversations with the user, emulating human-like conversations. In the healthcare space, emotionally intelligent voice-activated assistants are shown to be effective in addressing mental health problems, which will be explored in the subsequent chapter.

2.2 CURRENT APPLICATIONS OF VOICE-ACTIVATED ASSISTANTS

Analyzing the current applications of voice-activated assistants can give us valuable insight into the development of voice-activated smart hospital assistants. Advancements in the consumer sector hint at what is possible in the clinical and surgical settings.

2.2.1 Mobile Assistants

The most prevalent modality of voice-activated assistants is the mobile assistants that reside in smartphones. Siri, Google Assistant, and Bixby are part of this category. This is by far the most used modality of voice-activated smart assistants. Mobile assistants can carry out tasks within the phone ecosystem and can also access the phone's network, responding to questions and completing network-reliant tasks. Mobile smart assistants have been attributed to giving an identity and soul to smartphones, allowing for deeper connections with the user.

The introduction of Siri in 2011 significantly expedited the mainstream use of voice-activated smart assistants. Depicting Siri as simply another cool addition to the iPhone 4S would be an understatement. After seeing Apple's success with Siri, competitors followed suit, opening the age of mainstream mobile smart assistant technology.[3]

Today, mobile smart assistants play a vital role in the space of voice-activated smart assistant technology. Around 42% of adults in the US use voice-activated assistants on a smartphone, which makes up the highest percentage of all smart devices. We expect the mobile smart assistant space to continue expanding and serve as the voice of smartphone manufacturers, providing an important avenue for user connections.

2.2.2 Smart Home Assistants

After mobile assistants, the second most prevalent use of voice-activated assistants is smart home devices, which are placed in users' homes.

Home automation technology began to garner interest in the 1970s, as many dreamt of having their own personal assistants or live-in butlers. Growing affluence and a booming middle class after World War II led to the growing adoption of European luxury in American homes. Although the idea of having live-in butlers was considered the pinnacle of luxury for the wealthy, this was out of reach for the middle class. The gap was bridged in 1975 when the first home automation technology became available. The device, called X10, acted as a wired communication control hub that automatically controlled lights and electrical outlets. Today, home automation has taken huge new leaps, thanks to voice-activated smart assistants.[4]

Current applications of home automation technology allow everything from controlling home devices like lights and thermostats to providing home security. Smart home devices can also sync with other IoT devices and home appliances to provide interconnected convenience. Now, tech companies are branching out into the healthcare space to provide at-home care using smart home devices. Users can order prescription medications with smart home assistants and, in the near future, will be able to consult smart home devices for medical advice. Tech companies are also tackling at-home education, offering interactive teaching software for children and teens.[5,6]

New experimental work is being done to provide care for elderly and disabled persons, allowing for semi-supervised home care. Continued innovation will allow patients and their caretakers to have greater autonomy owing to smart home devices.[7,8]

2.2.3 Commercial Assistants

Voice-activated smart assistants are not confined to the consumer space. In fact, commercial applications of voice-activated smart assistants are disrupting many industries. At Rockwell Collins, engineers are developing voice-controlled airplane cockpits. Using just their voices, pilots can change the speed, altitude, and heading without exerting manual controls. Additionally, they can tune the radio or display important aviation charts. This technology has been shown to reduce 75% of the time to task completion and to enhance situational awareness and focus.[9]

The manufacturing industry has also seen the rise of voice-activated assistants. Companies like PTC are developing industrial automation for their manufacturing plants, allowing plant maintenance workers to monitor automation equipment and plant machinery.[10] GE also announced that it will equip many of its products with voice-assistant technology, from its jet engines to power plant turbines. GE industrial workers can now directly ask their jet engines about maintenance status. It is expected that by the end of 2022, more than half of industry 4.0 ecosystems will utilize voice-assistant technology. This technology will hopefully enhance human–machine interactions, allowing for greater volume, improved time optimization, and better quality control.[11,12]

Lastly, many companies are transitioning away from providing customer support through human specialists, opting instead for voice-activated smart assistants. Customers can now call company support to find a smart assistant replying on the other end of the call, listening to commands or questions to assist them. Fewer human customer specialists are needed than in the past, improving efficiency and cost for many companies.

2.3 THE FUTURE OF VOICE-ACTIVATED ASSISTANTS

Understanding the future trajectory of voice-activated assistants can advise our approach to apply this technology to the medical field.

2.3.1 Conversational AI

A key objective of voice-activated assistants is to reach a level where natural conversations with users are possible. Currently, voice-activated assistants have pre-set functions that are accessed through a central terminal. Commands and responses are hardcoded into the system, which prevents flexibility. The next phase of development should focus on making the most human-like voice-activated assistant. This means understanding the context and the nuances behind users' commands, allowing for easy follow-up. In essence, users will feel like they are talking to an actual person. In the near future, smart assistants will be able to make jokes, use sarcasm, and even understand slang. As smart assistants become more human-like, users will develop deeper emotional connections with them, which has great implications for psychiatric therapy. Smart voice-activated assistants could potentially provide therapy sessions, listening to users' concerns and addressing them accordingly based on the clinical practice guideline laid out by the American Psychiatric Association. For this to happen, natural language processing and artificial intelligence need to make a quantum leap in the coming decade.

2.4 CHALLENGES

2.4.1 Privacy

In 2017, Burger King released a 15-second Super Bowl ad featuring a Burger King employee who complains about the ad's short time limit, which prevents him from explaining the greatness of its burgers. In the last few seconds of the ad, the employee cleverly hands off the task to all the Google smart assistants passively listening.

"Okay Google, what is the Whopper Burger?"

Within seconds, Google Home devices across the US started reading out Wikipedia's definition of the Whopper Burger. Google took quick action to prevent further activation, but Burger King prepared several versions of the ad that circumvented Google's fix. Although it was deemed a clever marketing move, this advertisement fiasco sparked controversy, as it uncovered a fatal security flaw in voice-activated assistants – the lack of security measures allowed universal activation of voice assistants that have access to private personal data.[13]

Security and privacy are critical issues affecting electronics and networks, and voice-activated assistants are no exception. Big tech companies are adamant about making their voice-activated assistants free of security vulnerabilities, but breaches at the hardware and software levels were recently uncovered. A team of American and Chinese researchers discovered that ultrasonic waves, which are inaudible to the human ear, were shown to activate voice-activated assistants.[14] An ultrasound emitter placed in proximity of the smart assistant produced ultrasonic waves that activated the micro-electromechanical system embedded in the smart assistant. Precisely calibrated ultrasonic waves tricked the micro-electromechanical system device into transducing incoming mechanical vibrations to electrical signals. The researchers were able to successfully hack 15 of 17 smartphone voice-activated assistants. Another research group discovered a method to breach the security wall using just light. Sugawara and his team used oscillating lasers to emulate an audio signal, successfully hacking voice-activated assistants from 100 meters away.[15] These security vulnerabilities allow hackers to access private information, or even worse, home security – for example, opening garage doors. As the scope and capabilities of voice-activated assistants expand, so too must the security.

2.4.2 Public Perception

A major hurdle for voice-activated assistants is the public perception of voice AI. When it was first introduced to the public, voice-activated assistants were deemed avant-garde. As the hype started to fade and progress slowed, so fell the public perception. Despite the expansion of skill sets, basic tasks like setting an alarm make up the majority of use-cases. Most users have yet to advance to more sophisticated features like shopping or controlling other IoT devices. According to the Pew Research Center, one reason for the lack of advancement lies in the general mistrust of voice AI. Consumers simply cannot trust it to accomplish more complex tasks when its reliability to execute even the basic tasks is deemed fair at best. This has been a significant deterrent for users, and these concerns must be addressed.[16]

Many consumers are not comfortable using a voice-activated assistant in public due to its perception. The act of using voice AI in the open is often seen as an unnecessary gesture. More importantly, some consumers are hesitant to use verbal methods, as it may reveal personal information.[17]

2.4.3 Exclusion of Certain Users

One critical limitation of voice-activated assistants is that some are unable to use the technology. For instance, people with hearing disabilities, speech disorders, and language learning disabilities may not be able to make use of voice-activated assistants. Further innovation needs to take place to accommodate these subgroups.

2.4.4 Introduction to Healthcare Applications

So far, we have looked at voice-activated assistants that are widely used in the consumer sector.

However, the same technology that has existed for more than half a century has not meaningfully breached the healthcare space. Yet, it is only a matter of time before these advancements make their way to clinics around the world, and healthcare providers will soon team up with smart virtual assistants. Therefore, it is imperative to not only introduce the wide-ranging clinical applications of this technology but also to outline the appropriate steps to safely and correctly implement it for optimal healthcare delivery and efficiency.

The remainder of this book will explore the development of the smart surgical assistant (SSA) and review its latest deployment in the operating suite, discussing the feasibility of the technology with respect to safety, workflow, and surgeon compatibility. It will also lay the groundwork that would guide future iterations of virtual assistants, focusing on overcoming hurdles related to development.

We will conclude by discussing the value of translational sciences and the advances they bring to healthcare innovation. This analysis will be grounded in lessons learned from developing the SSA and the merits it brings to the medical community at large. It will also analyze the challenges of translational sciences with respect to engineering innovation and will propose ways to advance current efforts.

REFERENCES

1. Olmstead, Kenneth. "Nearly half of Americans use digital voice assistants, mostly on their smartphones." *Pew Research Center* 12 (2017).
2. Ruan, Sherry, et al. "Comparing speech and keyboard text entry for short messages in two languages on touchscreen phones." *Proceedings of the ACM on Interactive, Mobile, Wearable and Ubiquitous Technologies* 1.4 (2018): 1–23.
3. Gross, Doug. "Apple introduces Siri, Web freaks out." CNN.com (2011).
4. Rye, Dave. "My life at x10." X10 (USA) Inc., USA (1999).
5. Dojchinovski, Dimitri, Andrej Ilievski, and Marjan Gusev. "Interactive home healthcare system with integrated voice

assistant." *2019 42nd International Convention on Information and Communication Technology, Electronics and Microelectronics (MIPRO)*. IEEE, 2019.

6. Terzopoulos, George, and Maya Satratzemi. "Voice assistants and artificial intelligence in education." *Proceedings of the 9th Balkan Conference on Informatics*. Association for Computing Machinery, 2019.

7. Hajare, Raju, et al. "Design and development of voice activated intelligent system for elderly and physically challenged." *2016 International Conference on Electrical, Electronics, Communication, Computer and Optimization Techniques (ICEECCOT)*. IEEE, 2016.

8. Balasuriya, Saminda Sundeepa, et al. "Use of voice activated interfaces by people with intellectual disability." *Proceedings of the 30th Australian Conference on Computer-Human Interaction*. Association for Computing Machinery, 2018.

9. Trzos, Michal, et al. "Voice control in a real flight deck environment." *International Conference on Text, Speech, and Dialogue*. Springer, 2018.

10. Stackpole, B. "Are virtual assistants headed to the plant floor?" AutomationWorld.com (2020).

11. Gillies, C. "Talk! With! Me! Digital assistants are on the rise." Transformationbeats.com (2017).

12. Longo, Francesco, Letizia Nicoletti, and Antonio Padovano. "Smart operators in industry 4.0: A human-centered approach to enhance operators' capabilities and competencies within the new smart factory context." *Computers & Industrial Engineering* 113 (2017): 144–159.

13. Golgowski, Nina. "This burger king ad is trying to control your google home device." *Huffpost*, April 12 (2017): 7.

14. Yan, Qiben, et al. "Surfingattack: Interactive hidden attack on voice assistants using ultrasonic guided waves." *Network and Distributed Systems Security (NDSS) Symposium*. Institute of Electrical and Electronics Engineers, 2020.

15. Sugawara, Takeshi, et al. "Light commands:{Laser-based} audio injection attacks on {voice-controllable} systems." *29th USENIX Security Symposium (USENIX Security 20)*. Association for Computing Machinery, 2020.

16. Pfeifle, Anne. "Alexa, what should we do about privacy: Protecting privacy for users of voice-activated devices." *Washington Law Review* 93 (2018): 421.

17. Easwara Moorthy, Aarthi, and Kim-Phuong L. Vu. "Voice activated personal assistant: Acceptability of use in the public space." *International Conference on Human Interface and the Management of Information*. Springer, 2014.

DEVELOPMENT OF SMART SURGICAL ASSISTANTS

3.1 INTRODUCTION

Why are we focused on voice-activated assistants in the clinical space? Are they even viable solutions that could optimize healthcare delivery? To answer these questions, we must understand that the era of clinical AI is rapidly approaching. In fact, clinical AI is being implemented in clinics across the world as this book is being written. There are AI algorithms that detect cancer during colonoscopy.[1] And certain companies, like Jvion, optimize healthcare costs and patient outcomes by analyzing millions of pieces of patient and hospital data. In the not-so-distant future, human healthcare providers will work side by side with clinical AI to improve healthcare delivery for the better, providing preventative, personalized, and value-based care.

As mentioned previously, vocal communication is the most natural form of communication modalities for humans. Therefore, the most effective and intuitive way for clinicians to interact with AI is via vocal command. This is most certainly the case for clinicians in hectic hospital environments, where prompt and concise delivery is essential. Healthcare providers do not have time to sit and type out each and every command for executing clinical AI applications. Therefore, voice-activated hospital assistants will have a far-reaching role in interfacing human healthcare providers with clinical AI.

The surgical environment is the very first clinical department into which voice-assistant technology will be integrated. This choice was made with good reason: as mentioned in the previous chapter, the most highly cited factor that pushed consumers to adopt voice-activated assistants was the convenience, as they free up their hands and allow them to remain focused on the current task. Neurosurgeons, general surgeons, orthopedic surgeons, and members of 12 other major surgical specialties conduct lengthy operations that demand

DOI: 10.1201/9781003253341-3

meticulous hand–eye coordination. It was therefore apparent that surgeons would benefit most from adopting this technology.

3.1.1 Similar Technology at Work

Today, the use of voice AI in the clinic mainly occurs in the form of dictation services. Companies like Nuance and Suki provide products that allow healthcare professionals to record patient notes verbally. In certain situations and circumstances, healthcare professionals are required to document pertinent patient information quickly and concisely. For physicians, that includes writing medical records that detail patient-centered symptom descriptions. For medical scribes or documentation assistants, this means transcribing doctor and patient interactions to be stored in the electronic health record.

Nuance's Dragon Ambient eXperience or the Ambient Clinical Intelligence (ACI) service frees physicians from taking notes during patient appointments, allowing them to interact more personally with patients. The ACI works by simply transcribing the conversation between patient and physician. It also organizes the patient's symptoms and, using its AI algorithm, creates a differential diagnosis by contextualizing the conversation. This benefits physicians in two ways. First, patient interactions become more fluid. There are no pauses or interruptions between conversations as the provider jots down notes. This not only saves time but also allows patients to speak uninterrupted. Second, the ACI allows physicians to focus on the patient by eliminating the accessory task that dragged them down during these appointments. Pen and paper or monitor and keyboard are no longer the center of the interaction, and physicians are able to maintain good eye contact and attentive body language. Furthermore, patient information that was previously overlooked due to time and focus constraints can be given ample attention. Doctors can spend more time listening to patients' stories, backgrounds, and socioeconomic conditions and then can deliver a holistic treatment plan.

Nuance also works on Dragon Medical Virtual Assistant to reshape patient interactions. This software allows patients and clinicians to order prescriptions, schedule appointments, and create patient-specific notes. Clinicians can also search through the patient's medical history and conduct clinic-specific tasks, such as setting up appointment reminders. Suki operates in the same field and performs much the same function, allowing physicians to automate many administrative tasks. One notable highlight of Suki's software is that it allows physicians to easily draft patient medical history via voice. Suki can also navigate multiple directories within a patient file and append notes without having to type them out.

Currently, the smart voice-activated assistant technology available for clinics is confined to outpatient and at-home care. So far, we have not yet seen voice-activated assistants branch out to include department-specific functions – and rightfully so, as there are significant challenges associated with such deployment.

3.2 METHODS

3.2.1 Identifying Problems to Tackle

Before an SSA can be implemented in the operating room, we need to understand whether such technology has its place. Surveys of neurosurgeons and an extensive literature search were conducted to identify safety and logistical issues of the operating room. We found three problems that could potentially be addressed with the integration of the SSA technology.

3.2.1.1 Reduced Surgical Site Infection

Surgical site infection (SSI) is a postoperative infection on or near the surgical site. It is estimated that around 5% of patients develop SSI, even with strict precautions in place.[2] SSI occurs when the outermost layer of the skin and mucous membrane fails to prevent the entry of bacteria, viruses, and infectious agents into the body. As these protective layers are cut to expose the target tissues and organs during surgery, surgical sites become more susceptible to infections from the etiological agents in the environment. Therefore, except for rare instances of delayed closure, most surgical wounds are closed immediately to reduce SSI.

Surgical sites that require extensive healing are interrupted by SSI, which increases the likelihood of health complications and a costly reoperation. Symptoms of an SSI depend on its type and severity. Fever and germination of pus can result in a minor SSI, often leading to reopening of the wound site. In more severe cases, a culture of pus (i.e., an abscess) is formed and leads to blood or fluid drainage. Overall, the impact of SSIs can be mild or severe. Antibiotics are used for treatment, and surgery is needed to remove the infected tissue in severe cases.[3] These additional treatments delay recovery, increase cost, and compromise patient safety. Broex et al. found that patients with SSIs had a 34% to 226% increase in cost and a 48% to 310% increase in the length of stay at the hospital compared with patients without SSI.[4]

Due to the potential detrimental impact of SSI, certain precautions and protocols are set in place in the operating room. The crux of the protocol is a proper aseptic technique, which is emphasized in the operating room – no matter the procedure. In simple terms, proper aseptic technique emphasizes the need for sterilizing any tools or surfaces that may directly or indirectly come in contact with the surgical site. Proper hand washing, appropriate attire (i.e., surgical scrubs), and sterilization of surgical tools are all important preoperative preparations to control SSIs. In any operating room, the sterile field is clearly separated from the non-sterile field, and this boundary is often indicated by physical barriers. This division ensures that any tool or person entering the sterile zone is free of microorganisms that may introduce harmful infections.[5]

Even with these precautions in place, however, etiological agents can still breach the sterile field and infect the patient through contaminated surgical tools or surfaces. Numerous studies on bacterial transmission on major touch points in healthcare settings show that handheld instruments, computer peripherals, and improperly sterilized medical tools have significant etiological agents.[6,7] And these improperly sterilized touch points in the operating room can act as reservoirs of nosocomial agents. In a study on pathogen transmission in the operating room, the mean number of contamination sites was 31 sites across 10 simulated surgical procedures; and 13 of those sites were contaminated 100% of the time.[8] These studies illustrate that the operating room is not a bacteria-free environment and signal that cross-contamination through non-surgical accessory tools is a major issue.

3.2.1.2 Shorter Operative Time

Prolonged operative time is associated with suboptimal surgical outcomes, as many multivariate studies have shown. In a meta-analysis of prolonged operative time, researchers found a 14% increase in the likelihood of complications with every 30 minutes added to the operative time.[9] Also, a study on esophagectomy found that longer operative time was correlated with health complications like pneumonia, septic shock, unplanned reoperation, and mortality.[10] Similar health complications are also shown in prolonged laparoscopic procedures.[11,12] Although longer operative time can be a byproduct of a complicated procedure with greater health complications, reduced operative time still benefits patients in many ways. First, reduced operative time translates to a reduction of surgeon fatigue, thereby yielding greater focus and decreasing the probability of surgical mistakes.[13] Second, shorter procedures can reduce patient-related complications. Reduction of surgical time means that the wound is open for bacterial transmission for a smaller window – thereby reducing the chance of SSI.

The effect of reduced operative time on the healthcare system is immense. First, reduced operative time results in better patient turnover rates in the operating room, which increases the number of procedures conducted. Second, reduced operative time results in lower cost per surgery, greatly reducing healthcare costs for patients. According to a study published in *JAMA Surgery*, the cost of running an operating room in California is $36 to $37 per minute.[14] For a 6-hour surgery, that equates to approximately $13,000. If opportunity cost or missed treatment opportunities are added into the calculation, the financial burden rises even higher for the greater healthcare system.

3.2.1.3 More Efficient Allocation of Human Resources

A significant nursing shortage is looming over the US healthcare system, and healthcare needs grow each year. A study that investigated the nursing shortage in the US estimated a deficit of 510,394 nurses by 2030.[15] This growing concern must be addressed not only by advocating for increased enrollment of nursing students but also by effectively allocating nurses across hospital departments. New technology that can support current nurses, such as voice-activated smart assistants, can be introduced to increase the efficiency of patient care. If an AI voice-activated technology could support nurses by taking on some of the administrative and preparative work that occurs in the clinic, some nurses could be reallocated to short-staffed departments. The same can be said for operating rooms. If a voice-activated assistant could take on the role of one or two circulating nurses during surgery, this could facilitate the reallocation of nurses to other departments in need. Alternatively, it could result in a greater nursing turnover schedule, which in turn could decrease nurses' burnout rate.

Another benefit of reducing the number of personnel in the operating room is cost. Wages make up the greatest portion of the cost of running an operating room. In California, of the $36 to run the operating room per minute, $13 is spent on wages and benefits.

Reducing the number of operating room personnel without compromising patient safety could reduce the cost of surgery for patients. Furthermore, in studies that analyzed pathogen transmission in the operating room, it was found that overcrowded operating rooms have increased levels of airborne pathogens and infectious agents, which increases the risk of SSI.[16] Reducing the number of medical personnel in the operating room can therefore yield better patient postoperative outcomes.

3.2.2 Observations to Guide Our Proposed Solution

Before attempting to build an SSA, we took deliberate steps to analyze the current surgical environment. This analysis allowed us to brainstorm ideas and functions that could be utilized in our system. Neurosurgery cases were observed first, and we identified two main components that offered opportunities for improvement. First, we recorded workflow procedures in which the operating surgeon required the help of an individual outside the sterile field. Every instance was noted, and the verbal commands given by the operating surgeon were recorded. Second, we identified common non-surgical touch points in the operating room – both inside and outside the sterile field – to reduce SSI. The purpose of this process was to identify any possible contamination points that the SSA could eliminate by automating tasks that require non-surgical tools.

Our observations uncovered many interesting phenomena. We noticed that the operating surgeon requested the help of a circulating nurse primarily for non-surgical assistance. This included tasks such as retrieving patient data, adjusting the surgical bed and lights, and sterilizing tools or specimens. Another major task handled by the circulating nurse was managing equipment and supplies in the operating room; the circulating nurse frequently checked on the status and availability of these supplies. The majority of the circulating nurse's responsibilities were carried out preoperatively. The circulating nurse verified the identity of the patient and prepared the patient for surgery. Preparations include positioning the patient and readying their skin for incision by disinfecting the skin and removing excess hair. On rare occasions, the circulating nurse was asked to call a person outside of the operating room or was responsible for operating the camera for educational or research purposes.

As noted, our second aim was to mark any touch points in or out of the sterile field. A few of the touch points within the sterile field included sterilized operating room furniture, such as the surgical bed, electrosurgical units, suction system, patient monitor, and specialty and anesthesia equipment. Other touch points included the backtable and stands (e.g., mayo and basin ring stands). Outside of the sterile field, operating room furniture such as computers, linen and trash hampers, sponge/kick buckets, and sitting stools were present. A clear distinction of the sterile field was observed, and the effort to prevent contamination was obvious. It was evident from the observations that simple tasks that were easy and quick to complete by the traditional manual method should be left alone. Also, voice-control of complex electrosurgical units and specialty equipment was deemed out of scope for this project due to safety concerns. However, tasks that required many moving parts and personnel were considered to be potential areas of improvement. To reiterate from the previous section, traffic into, out of, and within the operating room should be heavily restricted to reduce airborne contaminants. Tasks like controlling the surgical bed and lights, using the computer to retrieve patient data, and controlling the surgical camera were deemed feasible tasks for a voice-activated assistant to perform. Replacing these tasks with a smart surgical system posed no apparent risk to patient safety and improved traffic flow and efficiency within the operating room (Figure 3.1).

3.2.3 Design Requirements

The end goal of this project was twofold. First, we wished to demonstrate a proof of concept; second, we aimed to carry out a feasibility study on implementing smart voice-assistant technology in the operating room, which could tackle existing conundrums in today's surgical environment. The design requirements of the SSA were chosen to align with this end goal in mind. The engineering of the hardware and software of the system therefore was simplified to reduce development time and to allocate greater resources to implementation and testing. Existing voice-activated assistant hardware and network solutions were also used to reduce development time.

3.2.3.1 System Design

The system, as shown in Figure 3.2, illustrates a multi-step process involving a few components, including the off-the-shelf smart home

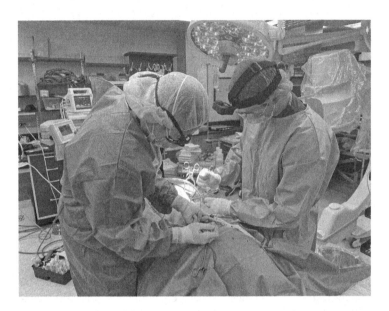

Figure 3.1 Two surgeons from Johns Hopkins Hospital conducting an animal surgery. This was part of the preliminary study to observe the workflow of a neurosurgery procedure. We noted the interaction between the surgeon and the circulating nurse as well as identified non-surgical touch points in the operating room. The sterile field is marked by the drapes, and the surgical equipment is visible in the background.

device. In the front end of the system, the smart home device collects the audio input from the user and uses natural speech recognition to translate the spoken words. The deciphered string is then fed to a third-party script, which controls the software or hardware that would execute the function. The development mainly involved customizing the smart home device to recognize SSA commands and programming the third-party script to receive and direct software and hardware components. One of the hardware elements that was built for this application was a voice-activated smart surgical bed that connected to the local network via Bluetooth. The development of the smart surgical bed will not be described here, as it would detract from the focus of the section; however, readers are welcome to explore its development cited in the list of references (Figure 3.2).[17]

The primary design requirement boiled down to choosing the best off-the-shelf smart home device for our needs. There were two key features that were crucial to be included in our system.

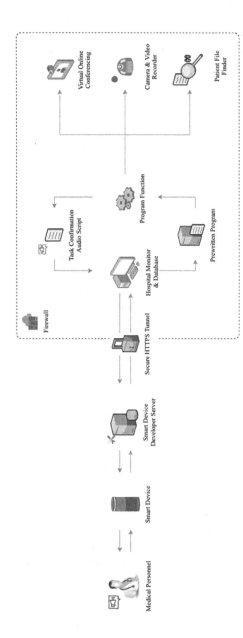

Figure 3.2 Process flow diagram of the Smart Surgical Assistant (SSA). The process begins when the operating surgeon states a command to the smart device, which delivers it to the smart device developer server. The developer server then establishes a connection with the custom server via a https tunnel, after which the custom server relays the command to a pre-written python script. This script will instruct the appropriate hardware and software components to execute the functions. Task completion notification is relayed back to the user.

First, in order to develop all of the chosen functions, the smart home device needed to allow incredible development flexibility. It had to possess an extensive developer server that would allow for the customization of commands. This feature allowed us to ensure that each command was executed with true natural speech recognition. It also required the flexibility to connect to external scripts that allowed the transmission of outputs and receipt of inputs. Because the control of external devices relies on these scripts, it was imperative that we could control these scripts through the developer server.

Second, the built-in voice-recognition system needed to perform well in a loud environment, such as an operating room. The smart home device had to be able to overcome the noise generated by surgical tools (e.g., surgical drills) and accurately capture the user's voice. The response time between commands also needed to be short to reduce lag time between commands. These criteria would not only enhance user experience but would also improve patient safety. Failure to recognize the commands due to noise or increased lag time could disrupt the surgical workflow and prolong the surgery.

Considering the limitations on the development time, cost, and design requirement, the following functions were carefully narrowed down. These included retrieving patient information (e.g., patient images and videos), adjusting the surgical bed and lights, managing communication with outside personnel, and recording the procedure.

3.2.4 Development of the System

With the specifications of our system outlined, we started to develop the system prototype. First, we began with customizing the developer server for the smart home device. We imported the verbal commands observed in the operating room to the respective functions that we wanted to execute. Some phrases that we included were "Raise the bed", "Take a picture", and "Open patient files". With these phrases, keywords were marked for the SSA to identify. In the example phrases, the keywords were "raise", "bed", "take", "picture", "open", and "files". It was imperative to include synonyms of the keywords to allow for natural speech. For instance, for the keyword "raise", we included the words "lift" and "increase". Having variations of these keywords allows surgeons to comfortably instruct the SSA to perform a task without memorizing a script of specific instructions. Surgeons can simply ask the SSA to perform a task just as they ask a circulating nurse.

Next, we hosted a locally secure web server on a computer to mimic the network hosted by an operating room computer. This web server was loaded with sample patient files, image-capturing program, video conferencing program, and a custom-written python script. The python script had three main components. First were pre-built libraries that allowed the script to take in the commands transmitted by the developer server. Second were separate functions for each of the designated tasks that instructed respective hardware and software components. These hardware and software include the microprocessor of the surgical bed, camera, and video conferencing application. Each function was linked to specific keywords predefined on the developer server mentioned previously. This ensured that only the requested function would be executed. Finally, each function had a feedback component that contained phrases that relayed back to the developer server to announce the status of the command performed. Having this feedback was crucial so that the surgeon was promptly informed of the status of the requested task. The feedback reported success, failure, or an error message.

Now that we had both sides of the system in place, we launched an active and secure https tunnel to allow for two-way communication between the developer server and our custom web server. Commands given from the developer server could reach the python scripts hosted on our web server by bypassing the firewall. Moreover, our web server was able to transmit feedback to the developer server without requiring extra steps.

To summarize, the surgeon's command is stored and processed as a JavaScript Object Notation (JSON) object. The developer console then matches it to the keywords and their respective variations mentioned above, and this information is sent to the custom web server via the secure https tunnel, allowing it to bypass the firewall. Next, the preexisting libraries loaded on our custom python script convert the JSON object into a compatible file that can execute the respective functions in the python script. The script then instructs the right hardware components and software programs to execute the task. Finally, after the task is completed, the python script sends one of the feedback responses to the developer server, which announces it to the surgeon.

The surgeon needs to follow a specific workflow to use the system properly. First, the surgeon activates our system by calling out the designated wake word. This wake word activates the SSA and instructs it to begin listening for commands. For Amazon Alexa and

Google Home, the respective wake words would be "Hey Alexa" and "Okay, Google". Once the wake word is recognized, the surgeon can begin instructing the SSA. Within each of the main commands, we built in sub-commands that allow the surgeon to instruct until they exit the command loop. For instance, the surgeon can command the SSA to open a patient PET scan and have it displayed on a screen. Before exiting this main command, the surgeon can ask the system to show the next scan by saying "next". This is one of many sub-commands embedded in the main command. The surgeon can exit the command loop by calling out a wake word and instructing another main command. Having these sub-commands in place saves time by not having to reinitiate the same main command *over and over again*. If there is a pause greater than a minute, the SSA would assume that the surgeon has finished and exit the loop.

3.2.5 Testing and Implementation

After the development came the implementation of the system. A surgical simulation center was used for this step. Before we installed our system, the operating room furniture was repositioned to mimic a neurosurgical operating room. Then the hardware, including the smart home device, surgical bed, and a laptop, was installed. The smart home device was placed in close proximity to the operating surgeon – directly in the line of his vocal projections but outside the sterile field. The laptop was placed next to the surgical displays with easy access to the circulating nurse.

Two neurosurgery residents from Johns Hopkins Hospital were asked to participate in the simulation. Each resident was paired up with a circulating nurse to perform a mock laminectomy procedure on a mannequin. First, they were asked to perform the surgery using the traditional method, whereby they instructed a circulating nurse to complete non-surgical assistive duties. Following the initial surgery, they performed the same procedure but using our SSA. They were asked to repeat the procedure again two weeks later, but with the SSA system tested first, followed by the traditional method. This was meant to eliminate the confounding variable of performance enhancement resulting from procedure order arrangement. Total surgery time was measured in each trial, along with the time to complete individual non-surgical assistive tasks by the circulating nurse and by the SSA. During the simulation trials, the circulating nurse was asked to put on gloves dipped in Glo-germ. Any surfaces touched would

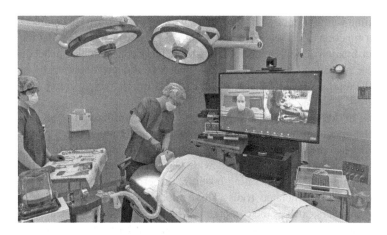

Figure 3.3 Mock simulation of the Smart Surgical Assistant display-ing the successful demonstration of the video conference feature. On the left bottom corner are the anesthesia machine and a cart containing surgical tools. Directly under the surgical monitor sits the smart device. Behind the monitor is a computer with a built-in custom server. The surgeon is operating on a mannequin and instructing the smart surgical assistant via voice. The circulating nurse is shown on the right.

leave a trace of the substance, later visualized by ultraviolet light. After the surgery, the residents were asked to provide their qualitative feedback on three criteria: level of difficulty, comfort, and feasibility of the SSA (Figure 3.3).

3.3 RESULTS

In these simulations, the SSA improved operating room efficiency and workflow. There were no interruptions that distracted the focus away from the surgery. The residents were also able to instruct tasks in a flexible manner, using the vocabulary they preferred. This allowed for a low learning curve and intuitive handling. Additionally, all commands given during the simulated surgeries were accurately recognized by SSA, with no errors or misunderstandings.

In terms of quantitative performance, the SSA saved a significant amount of time compared with the traditional surgical method. The simulated surgeries conducted with a circulating nurse took an aver-age of 47 minutes and 31 seconds. Additionally, it took approximately 98 seconds for a command to be instructed and be completed by the nurse. However, for the trials that utilized the SSA, the total surgical time was 43 minutes and 2 seconds, with an average time to complete a function of 57 seconds. The results of these trials showed that the utilization of SSA in the operating room has a considerable impact on the reduction of the total surgical time. SSA reduced about 9.43% of the total surgical time and the time to perform individual commands by 41.84%.

In addition to measuring the time saved by utilizing the SSA in the operating room, we found that there were 11 major touch points that had the potential for bacterial germ transfer. Again, these major touch points were visualized with Glo-germ. The majority of the touch points were operating furniture or peripherals used frequently during the simulated trials. These touch points included physical controllers, such as the computer keyboard and mouse, surgical light controllers, and handles attached to surgical carts. When comparing to touch points of the simulations that did not use SSA to those that did, we found that incorporating SSA into the surgical workflow eliminated 4 of the 11 major touch points. These were the computer keyboard and mouse, surgical bed controller, and surgical light controllers (Figure 3.4).[18]

3.4 DISCUSSION

When asked to provide feedback, the residents commented most frequently on the intuitive workflow of the SSA, emphasizing the convenience and flexibility of using verbal commands. Because they were already used to giving vocal commands, they found the transition to the SSA to be straightforward. They also noted that using the SSA did not create any disturbances or interruptions in the surgical workflow. Furthermore, the reduction of surgical time was greatly appreciated. They reported that the flexible platform and convenient user interface had major potential in the operating theatre.

One major criticism was the need to memorize the command structure of the SSA. The residents noted that recalling the main and subcommand layouts took minor focus away from the surgery. Although the surgical residents successfully recalled all instruction phrases, any failed attempt to deliver the message to the SSA had the potential to slow down the surgery. They also noted that it was difficult to move back and forth between main commands. When the residents needed to refer to patient images during a video conference call, they had to exit the command loop and reinitiate the patient file finder. Future development should focus on improving the transitions

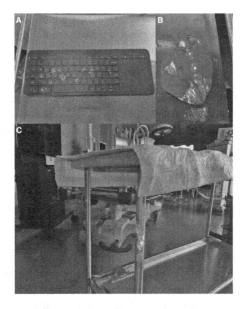

Figure 3.4 Photos of surgical equipment displaying touch points. UV light is illuminated on the surface, which visualizes possible contamination points. The top figure shows a keyboard used to operate the surgical monitor. The use of the smart surgical assistant eliminated the use of the keyboard, removing a potential nosocomial site. The bottom figure shows a surgical cart situated outside the sterile field. This is one example of a touch point that the SSA could not eliminate. Future development should focus on eliminating major touch points like the surgical cart handle by automating non-surgical processes using voice technology and robotics.

between the commands. Also, future versions of the SSA should allow for one or more commands to function passively when other commands are active.

The demonstrations showed that there is clinical efficacy of the SSA. We hypothesized at the beginning of this book that voice-activated assistants will be a useful user interface for surgeons whose hands are mostly occupied during surgery. The feedback provided by the residents validated our hypothesis. The SSA was not only well received but was also effective in addressing the clinical problems we set out to solve. First, the time to complete surgery was decreased as a result of the reduction of time to complete non-surgical assistive tasks. As subsequent improvements are made to the system, we believe that the SSA will be able to save even more surgery time in the future. Second, the SSA was successful in emulating the tasks traditionally performed by circulating nurses. With more features added to the system, the SSA may be able to ameliorate the impending nursing shortage by taking an active role as surgical assistants. Lastly, the SSA played a significant role in reducing the number of touch points in the operating room, thereby reducing possible contamination or SSIs. Critics may point out that the contamination points outside the sterile field are irrelevant to SSIs. However, as the literature has shown, it is better to eliminate any and all germs in the operating room due to the possibility of indirect contamination resulting from human error. Elimination of the keyboard and mouse from the touch points was a significant step in the right direction, mainly due to the fact that they are one of the most contaminated surfaces in the operating room. Again, as the SSA expands its features to connect surgical tools and other operating room furniture, we can conceive a fully automated and touchless surgical theatre in the near future.

On the subject of improvements, a few need to be addressed. During the development of the SSA, we realized the need for greater emphasis on security on both the front and back ends. On the front end, the system should limit access by unauthorized persons. This security measure could be accomplished by adding an authentication method to validate the user's identity (e.g., voice authentication) before patient data can be accessed. Depending on the emerging HIPAA compliance on voice-activated smart assistants, patient data access via the SSA may be somewhat limited. On the back end, developers should concentrate their efforts on methods to incorporate the system into existing hospital networks. Will the SSA have access to the hospital intranet? What additional security measures are needed to block unwanted network access or activity? How will the SSA access electronic medical records and picture archiving and communication systems? These are questions that future developers should focus on and tackle before an SSA can be deployed.

3.4.1 Merits

Merits of the SSA were demonstrated during testing and implementation. First, the SSA significantly reduced the time to complete non-surgical assistive tasks, which in turn reduced total surgery time. This reduction in surgical time alone has significant positive implications for patient outcomes, as lower surgical time is correlated to positive surgical outcomes. In future versions of the SSA, we expect a greater reduction in surgical time as a result of faster processing power and software optimization. If the reduction of surgical time is significant, hospitals could schedule more surgeries per day, which also improves health outcomes for the greater population.

Second, the SSA proficiently and autonomously carried out non-surgical assistive tasks once directed to do so. This is significant because it independently executed tasks traditionally assigned to one or two circulating nurses. As observed during testing, the SSA was able to free up a circulating nurse who could then concentrate on more important surgical assistant duties. In future operating rooms, we can safely predict that fewer medical assistant personnel would be required as more surgical and non-surgical assistive tasks are handled by AI, ultimately reducing cost in the long term. This can also be used to address a nurse shortage by reallocating staff to other departments in need.

The feasibility test also validated that the use of vocal commands is a natural fit in the surgical environment. We hypothesized that the operating surgeon, whose hands are usually tied up, would find voice commands to be useful. When compared to the gesture user interface, a non-traditional surgeon control unit that relies mostly on finger and hand gestures, we found that voice command was superior in many categories. First and foremost, the surgeons' use of vocal commands freed their hands, allowing them to focus on the surgery. This would not be the case in the gesture user interface, as the surgeon needs to remove their hands from the operative field to execute control over a surgical apparatus, hindering workflow and interrupting concentration. One important criticism of the gesture user interface that has yet to be addressed is its susceptibility to cross-contamination. Surgeons whose gloves are contaminated with patients' specimens or fluids could be dispersed to nearby surgical tools or furniture when performing gestures. With the SSA, the surgeon could instruct commands without worrying about the spread of fluids.

All in all, the feasibility test successfully laid the foundation for future smart device developments and deployments. Positive feedback from surgeons, as well as successful study results, warrant continued development of this system. This study, in essence, demonstrated that the clinical voice-activated smart assistants are not limited to the current boundaries of medical record keeping but can extend their scope to the operating room, playing an active role in assisting the peripheral tasks of surgeries.

3.4.2 Future Outlook

Where does the SSA go from here? Based on this successful demonstration, we are optimistic that research groups and companies will see the unfulfilled potential of smart assistants in the surgical setting. With more active research and development dedicated to such endeavors, we will likely see rapid commercialization and deployment of the SSA in operating rooms across the country in the not-so-distant future. Nevertheless, significant barriers to innovation are present, and companies and research groups will be obligated to follow strict medical sector regulations. We also expect some deterrence and backlash from the medical community at large due to the conservatism and the perceived implication this system has on human resources.

In terms of future SSA development, three short-term improvements should be addressed. First, we see that more features can be added to make the system more comprehensive. This can be accomplished by linking the system to other surgical apparatus and tools, allowing the SSA to have greater control. Although this may sound far-fetched, it is actually close to reality. Voice-activated surgical tool control was accomplished decades ago with the ZEUS Robotic Surgical System, which used a voice-controlled robotic arm for precise endoscope control.[19] The adoption of voice-activated surgical tools can start with monitoring surgical tools of their working status. However, careful selection of surgical tools and substantial research needs to be conducted to validate the voice-control of surgical tools.

The long-term outlook of the SSA is exciting, with the concurrent development of AI and smart assistant technology. The SSA would greatly benefit from the increased natural-language processing and the capability to understand the context and intent of the user. This is a criticism not only of our system but also of other advanced smart home devices. Current smart assistants are unidimensional; unlike humans, they lack the understanding of context. The absence of memory to store recent conversations and the inability to form connections between them make it impossible for users to have uninterrupted and fluid conversations. In our system, we found that the surgeons needed to call on the SSA each time they wanted it to perform a main task, breaking the workflow. For instance, if a surgeon asked the SSA to open a patient's CT scan but realized he needed the MRI instead, the surgeon would have to initiate another command from the beginning instead of saying, "No, not that one. I need the MRI". A human assistant would perfectly understand the surgeon's command because they do not need to extrapolate further content and intent of the surgeon. The SSA and the smart home device lack extrapolation and need to be spoon-fed with commands. This problem is not unique to the SSA. When one asks about the temperature outside, one cannot follow up with a question of "What is that in Celsius?" without receiving an error message from the smart home device. Though they may reliably perform tasks when given adequate context, the current state of the technology limits the SSA from reaching the level of human comprehension and execution.

We believe that AI and machine learning can enhance the working capabilities of the SSA on many fronts. If the SSA is able to recognize a pattern and predict the commands that it would likely be given, it could provide the user with suggestions even before the commands are spoken. We see how valuable this is in our own personal lives. Google's email system, Gmail, incorporates Smart Compose, which offers word or phrase suggestions, allowing users to write faster. Google Maps also uses machine learning to predict traffic based on historical traffic patterns on roads over time, informing the user of the best route or possible delays before they embark on a journey. Google Maps can also predict our destination at a certain time of day based on our previous location data. With the tap of a button, Google Maps guides users without needing them to manually type in the destination. Given enough data, the SSA can also use predictive tools to assist surgeons based on the patterns learned in previous surgeries. If certain tools and surgical settings are consistent in specific procedures, the SSA can offer suggestions to the operating room staff, enhancing efficiency and optimizing user interaction. For instance, the SSA could learn that a Trendelenburg bed position is mostly used for colorectal or genitourinary surgery and could offer to change the bed position before a command is explicitly given. For the surgical personnel, these predictive suggestions can save time and reduce manual exertion of activating the SSA. The operating theatre will become smarter, faster, and capable of swift automation.

As hinted earlier, the SSA has opportunities to integrate with surgical tools. A new era of surgical interfaces is on the horizon, and when paired up with the SSA, these devices and interfaces can yield a powerful arsenal for surgeons. One such tool is the surgical virtual and augmented reality technology that projects images onto the surgical goggles or surgical space. SentiAR is one such company developing holographic augmented reality that projects a patient's anatomy above the patient as a reference check. Integrating a voice-activated assistant into the devices can elevate the surgeon–computer interaction to the next level by eliminating manual exertion of controlling augmented and virtual reality settings.

REFERENCES

1. Wang, Pu, et al. "Development and validation of a deep-learning algorithm for the detection of polyps during colonoscopy." *Nature Biomedical Engineering* 2.10 (2018): 741–748.
2. Khuri, Shukri F., et al. "The national veterans administration surgical risk study: Risk adjustment for the comparative assessment of the quality of surgical care." *Journal of the American College of Surgeons* 180.5 (1995): 519–531.
3. "Surgical site infections." Johns Hopkins Medicine, https://www .hopkinsmedicine.org/health/conditions-and-diseases/surgical-site -infections.
4. Broex, E. C. J., et al. "Surgical site infections: how high are the costs?" *Journal of Hospital Infection* 72.3 (2009): 193–201.
5. Berríos-Torres, Sandra I., et al. "Centers for disease control and prevention guideline for the prevention of surgical site infection, 2017." *JAMA Surgery* 152.8 (2017): 784–791.
6. Bures, Sergio, et al. "Computer keyboards and faucet handles as reservoirs of nosocomial pathogens in the intensive care unit." *American Journal of Infection Control* 28.6 (2000): 465–471.
7. Weber, David J., Deverick Anderson, and William A. Rutala. "The role of the surface environment in healthcare-associated infections." *Current Opinion in Infectious Diseases* 26.4 (2013): 338–344.
8. Birnbach, David J., et al. "The use of a novel technology to study dynamics of pathogen transmission in the operating room." *Anesthesia & Analgesia* 120.4 (2015): 844–847.
9. Cheng, Hang, et al. "Prolonged operative duration is associated with complications: A systematic review and meta-analysis." *Journal of Surgical Research* 229 (2018): 134–144.
10. Valsangkar, Nakul, et al. "Operative time in esophagectomy: Does it affect outcomes?" *Surgery* 164.4 (2018): 866–871.
11. Zdichavsky, Marty, et al. "Impact of risk factors for prolonged operative time in laparoscopic cholecystectomy." *European Journal of Gastroenterology & Hepatology* 24.9 (2012): 1033–1038.
12. Jackson, Timothy D., et al. "Does speed matter? The impact of operative time on outcome in laparoscopic surgery." *Surgical Endoscopy* 25.7 (2011): 2288–2295.

13. Slack, P. S., et al. "The effect of operating time on surgeons' muscular fatigue." *The Annals of The Royal College of Surgeons of England* 90.8 (2008): 651–657.
14. Childers, Christopher P., and Melinda Maggard-Gibbons. "Understanding costs of care in the operating room." *JAMA Surgery* 153.4 (2018): e176233–e176233.
15. Juraschek, Stephen P., et al. "United States registered nurse workforce report card and shortage forecast." *American Journal of Medical Quality* 27.3 (2012): 241–249.
16. Andersson, Annette Erichsen, et al. "Traffic flow in the operating room: An explorative and descriptive study on air quality during orthopedic trauma implant surgery." *American Journal of Infection Control* 40.8 (2012): 750–755.
17. Kim, Jeong Hun, et al. "Development of voice-controlled smart surgical bed." *Frontiers in Biomedical Devices.* Vol. 83549. American Society of Mechanical Engineers, 2020.
18. Allaf, M. E., et al. "Laparoscopic visual field." *Surgical Endoscopy* 12.12 (1998): 1415–1418.
19. Kim, Jeong Hun, et al. "Development of a smart hospital assistant: Integrating artificial intelligence and a voice-user interface for improved surgical outcomes." *Medical Imaging 2021: Imaging Informatics for Healthcare, Research, and Applications.* Vol. 11601. International Society for Optics and Photonics, 2021.

MERITS AND CHALLENGES OF TRANSLATIONAL SCIENCES

4.1 INTRODUCTION

The SSA is a part of a greater mission to improve healthcare outcomes and bring technological innovation to the field of medicine. Although our venture may be novel, the underlying theme of which it is a part is far from new. The act of applying existing engineering solutions to medicine is referred to as *translational science.* We have brought a solution that exists out in the consumer sector, proven through years of trial and error, and applied it to medicine. In this chapter, we describe the merits and challenges that come with this application, lay out our experiences, and set a call to action to expand our efforts in this space.

Before we embark on this step, we must understand the complex nature of translational sciences. The word "translation" comes from the Latin word for "carrying over", and in the field of medicine, translational science is the act of translating or applying scientific discoveries seen in the laboratory to patient care, ultimately improving healthcare delivery. We can broaden the definition, however. Translational science is the act of applying *any and all* observations in this world to medicine, with the goal of improving patient outcomes. It is another avenue taken by scientists, engineers, public health advocates, and medical doctors to solve a medical problem or disease by applying learnings from existing scientific, engineering, or social phenomena.

It is crucial to understand why this pursuit exists, to begin with. Our world is governed by the natural laws of physics, chemistry, and biology. The human body is also governed by these laws, but whereas we can drop a ball from a high altitude to test the laws of gravity, our bodies obviously cannot be the subject of these same experiments. Our bodies are too delicate and valuable; thus, human experiments must be done with caution and care. This is the very reason that any

DOI: 10.1201/9781003253341-4

drug or medical device pursuit involving a human trial is a slow and delicate process. We simply cannot test novel drugs without comprehending the risks behind them. To aid in this process, scientists came up with translational sciences. If it is inappropriate to first test a drug on patients, we can first do so in the laboratory, culturing cells or using microfluidic devices that mimic our bodily systems to test safety. We can conduct animal trials and see whether a novel medical device is producing an intended effect. We can precisely construct a solution and perfect it using a series of experiments and optimizations. Having performed this due diligence, we can then translate our findings to human clinical research and see whether it yields the intended effect. Translational science is ultimately a calculated approach to downsize the risk of negative outcomes and to greatly optimize for successful clinical implementation.

Another benefit of translational science is that it encourages continuous innovation and growth in the medical field. It pushes medical professionals to look outside of medicine for potential catalysts of innovation and change. This is an effective approach because the world outside of medicine changes more rapidly than it does inside. Other industries are regulated less stringently, so change is more rampant. New life-changing technology is invented every day, and the companies and consumers driven by cost and profitability become the final judges for a new technology's existence in the market. The healthcare field, on the other hand, plays by a different set of rules. The practice of medicine is evidence-based. Unless a change of practice is warranted by evidence of better patient outcomes and systemic improvement, it is not implemented. Here, the final judges are not the companies or the consumers (i.e., patients) but regulatory bodies and medical professionals. The answer must be "yes" to three important questions. First, is the new agent of change safe for patients? Second, does it work as intended? Lastly, is it better than the existing solutions currently implemented? Developing a new drug or device that upholds patient safety while outperforming current solutions is a slow, arduous process. Translational science, however, tries to tip the balance. It provides the means for new technological advancements made in other fields to seep into medicine. It provides the framework and the support for this endeavor and ultimately brings about innovation that is desperately needed.

The translational science of drug and device development is generally divided into five stages (though some institutes stage development differently), and each stage has a unique objective. The translation of medical technology is described in greater detail in Section 4.2. The first stage is the T0 stage, in which basic science research is conducted to lay the scientific foundation of the development. In this stage, scientists aim to define the treatment mechanisms, targets, and the chemical compound of an investigational drug. They also conduct necessary tests (e.g., animal studies) to demonstrate drug safety and efficacy. The second stage is the T1 stage, and this is the translation between basic science research and human clinical research. This stage aims to bridge the gap that exists between basic science and human clinical research by questioning the safety and efficacy of laboratory findings yielded and juxtaposing them with the biochemical mechanisms in human pathophysiology. Phase 1 clinical trials are performed at this stage, and these trials seek to understand the toxicity and the maximum tolerated dose of a drug. For medical device development, this step would evaluate safety and performance. The third stage is the T2 stage, and it sits between human clinical research and patient research. After validating that the investigational drug met the baseline safety requirement, the next step is to evaluate the efficacy of the drug. Phase 2 and 3 clinical trials are conducted in this step, focusing on patient outcomes with and without the investigational drug. In simple terms, this stage aims to discover whether the drug works as intended, by testing it on patients having the disease in question.

The penultimate stage of translational science is T3 and is often referred to as dissemination and implementation science. This stage aims to reduce the difficulties in implementing the outcomes of clinical studies into actual practice. If an investigational drug showed safety in human clinical trials, the next challenge is to take that investigational drug and fit it into clinical practice. How can we effectively disseminate the new investigational drug across clinics? For optimal drug interaction, what element in the drug delivery method should be kept constant? What could be changed to tailor it to different populations and communities? These questions need to be answered in this step for proper clinical implementation. The last step of translation is the T4 stage, and it occurs between practice of care to population studies. This final translation step attempts to analyze the effects of new drugs or devices on the population. At this point, researchers and regulators would look at the benefits and risks associated with the drug or device in question by studying health outcomes at the community level. This translational step would also entail changing public policies and community-level logistics to improve health outcomes[1] (Figure 4.1).

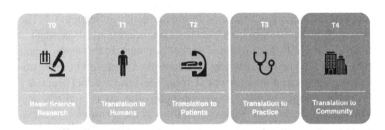

Figure 4.1 Translational science process flow diagram. T0 stage of translational science focuses on basic science research to lay the foundation of drug discovery. This step involves studying the mechanism of action and basic drug toxicity by conducting animal studies or in vitro tests. The T1 stage is the first translational step from bench top to bedside, and it investigates the safety profile of the investigational drug. Toxicity levels and maximum tolerated dose are studied. The T2 stage is the translational step for patients, and it examines the drug's efficacy on the intended patient population. The T3 stage is the translational step to guide the dissemination method. It investigates the best method to deliver the working drug to the intended patient population and tailors it to fit the community needs. The last stage or the T4 stage analyzes the impact of the drug on the community at large to produce public policies and community-level interventions to improve societal health outcomes.

4.1.1 Translational Science from the Engineering Perspective

We have described translational science from the perspective of drug development. But does the translation of medical technology follow the same process? The overarching steps are similar, but there are differences between the two. Note that the term *medical technology* here does not refer to the traditional term used for medical devices. Rather, it is based on the rising field of digital healthcare, a broad multidisciplinary field interconnecting medicine and technology. The main difference in the translational science process between drug discovery and medical technology lies in the initial design process and the first translational T0 step. The safety validation involved in the T0 step is significantly shorter for the medical technology development than it is in the drug discovery process. Rightfully so, the risk of implementing a medical technology is far lower than drugs. An investigational drug must demonstrate low cytotoxicity, hepatotoxicity, reproductive toxicity, and genotoxicity to account for patient safety in both the short and long term. Significantly fewer tests are needed to show the safety of medical technology. This means that the path to implementation may be significantly shorter for medical technology than it is for drug development.

The second difference between the two development courses is the design process. The procedure for developing a drug candidate takes on very few revisions once a chemical compound is designated for testing. Once a chemical compound is chosen, the drug is carried through the translational steps until it encounters an insurmountable barrier, which then brings the development back to square one. The barrier can be a result of significant toxicity or lack of drug efficacy. Medical technology, on the other hand, utilizes the engineering approach of problem solving, which allows constant revisions and debugging throughout the design process. For brief explanation, the engineering process is divided into six steps: idea, concept, planning, design, development, and launch. Engineers develop an idea taking into account existing solutions, requirements, and constraints. Then a specific plan is made to define milestones and schedule. During the design process, engineers sketch a drawing or a schematic to visualize the solution and realize it through the subsequent development process, during which prototypes are made. After validation and verifications, the product is launched. Even after the official design process is complete, engineers can revert back and make changes to the product as they see fit. Therefore, a solution continuously evolves over the design and testing cycle and adapts to hurdles presented along the way. The engineering method also allows for multiple prototypes for testing, greatly increasing the chances of successful implementation. For the development of the SSA, the team successfully utilized the engineering method to its full benefit. During the development phase, we engineered two separate working software models for two different smart home devices. During this same phase, we tested these two models by simulating a mock surgery in the laboratory to narrow down to the final candidate.

One unique aspect of medical technology development is the level of forecasting it allows for the developers. The engineering method in essence is designed to minimize unanticipated outcomes. The concept and planning stages exist so that the outcome is achievable. That is why an engineering project delivers a solution at the end more often than not – despite enormous challenges faced during development. Though the solution may not be perfect or faithful to the original design, a product still results. Drug discovery, however, is practically on the opposite end of the spectrum. Despite the countless hours spent on basic science research needed for compound development, only 7% of drugs that enter Phase 1 clinical trials make it all the way to approval and translation to the community.[2]

With a solution already in mind, engineers can simulate the workings of the solution in their heads, anticipating how it will play out in its environment. This level of forecasting offers a unique advantage for medical technology development: the ability to communicate with potential customers and users very early in the development cycle to identify their needs. Before commencing the development of the SSA, we were able to discuss our solution with surgeons and circulating nurses. As noted earlier, extensive interviews were conducted to address problems and inefficiencies in the operating room. We were also able to observe surgical procedures to get a better idea of the existing workflow and processes. Even before developing the prototype, we were thinking well ahead into the T2 and T3 stages of the translational science process, evaluating the risk factors associated with the SSA and planning how to minimize difficulties that would come with implementation. For instance, we were able to visualize the exact placement of the hardware in the operating room during the design process. This planned placement shaped the microphone settings and input volume of the system. The drug discovery process rarely grants this level of privilege in which the development team can foresee details of the implementation stage.

4.2 MERITS OF TRANSLATIONAL SCIENCE

4.2.1 Healthcare Innovations from a Solution-Oriented Approach

One of the main benefits of translational science is the medical innovations that stem from a solution-oriented problem-solving approach. The traditional approach to tackling a medical problem begins with an extensive study of the problem before solutions are considered. If we were to use this traditional approach, we would look for a problem in the healthcare field – for example, drug non-adherence. We would dissect the problem by analyzing individual substituents and devise solutions that account for these intricacies. The solution-oriented approach is a method often used by engineers and would approach this problem backward. A researcher using the solution-oriented approach would encounter an engineering solution, like block-chain technology, and customize it to tackle clinically relevant problems.

Imagine you were tasked with trimming the size of a forest. Instead of trying to tackle every tree species with a different set of tools, you would only bring one specific tool and focus on trees that this tool cut best. This is a larger-scale thought process that often goes unappreciated, even though a number of valuable medical devices and medical technologies have emerged from this process.

For instance, take the first implantable cardiac pacemaker. This device was engineered by Wilson Greatbatch, an assistant professor of electrical engineering at the University of Buffalo in 1956. Greatbatch was interested in recording fast heart sounds with an oscilloscope. He generated a testable signal using a 1-kilohertz marker oscillator with a 10-kilohms bias resistor. One day, he mistook a 1-megohm resistor for a 1-kiloohms resistor and placed it in the circuit. After turning on the oscillator, he saw that his circuit was generating oscillating bursts of a 1.8-millisecond pulse followed by a 1-second quiescent interval, greatly resembling the electrical activity of the heart. Greatbatch recounts that he stared at this phenomenon for a long while. He then recalled his lunchtimes spent with two New England brain surgeons a few years earlier, when they discussed the problem of heart failure in their patients. It then dawned on him that he could engineer a device based on the circuit he had inadvertently created to control the cardiac rhythm. Instead of brushing off his finding as a mere mistake, he used the solution-oriented problem-solving approach, working his way backward to identify a problem he could fix with his unintended discovery.[3]

The SSA is another device that used the solution-oriented problem-solving approach. Voice-activated assistant technology was actively being developed in the commercial industry for decades. We took notice of this technology, which brought immeasurable utility to everyday consumers, but that did not have the same impact on healthcare. We decided to use the solution-oriented problem-solving approach to spot the clinical problems that this technology could solve. We decided to integrate it into the operating room for reasons already discussed. However, before we embarked on this path we studied every clinical environment in which this technology had the potential to improve healthcare delivery and patient outcomes. This method of development was convenient and yielded much faster results than the traditional approach would have. Instead of limiting ourselves to one specific medical problem and conjuring all sorts of potential solutions to develop and test, we locked in on a solution that was within the domain of our expertise. This shaved off significant development and testing time and eliminated the possibility of not finding a solution. If we limited ourselves to one clinical problem, there is a good chance that we would have arrived at a solution that yielded no clinically relevant benefits. Then we would have to go back to the drawing board and start over. Let's go back to our scenario of trimming down the forest. What happens if we encounter a tree that cannot be brought down with the tools we have on our hands? It would be far more efficient to move on to the next tree than to devise a new solution to bring it down.

To summarize, the biggest merit of translational science is the catalyst of innovation in the medical field, as it encourages people to disrupt the status quo and to take an idea outside of medicine and ask, "What if?". This mindset encourages people to see the bigger picture and to pursue improvements in healthcare delivery. Again, innovation in medicine often results from innovation outside of medicine. Therefore, translational science asks us to never be complacent with the current implementations and practice of medicine. It asks us to always look for improvements. It asks us to be inspired by disrupting technologies outside of medicine and carefully bring those inside to change patient outcomes for the better.

4.2.2 Drug Device and Discovery

It goes without saying that the direct reward of translational science is its impact on the discovery of pharmaceutical drugs and medical devices. It is important to note how the majority of these drugs and devices are made. Whether it be an accident or deliberate endeavor, translational science has had an impact on the discovery of drugs and devices in one form or another.

Many of the most notable and impactful medical discoveries were made quite by accident. In 1895, Professor Wilhelm Roentgen from Wurzburg, Bavaria, was testing whether cathode rays could pass through a glass pane. While conducting his experiments, he noticed an interesting phenomenon when covering a cathode tube with black cardboard. Green incandescent light escaped and projected onto a nearby fluorescent screen. After more experiments, he discovered that this unknown light ray would pass through most objects, leaving a shadow of its internal structure. Professor Roentgen could have stopped there, put a name to this phenomenon, and called it a day. He would become the father of the x-ray and appear in all of our physics textbooks. But there is a reason his name also appears in our medical textbooks: he took the next step in asking how this new phenomenon had clinical relevance. He questioned whether our limbs were subject to this same physical phenomenon and illuminated his hand on the cathode tube. There he saw the silhouette of his bones and the internal tissues that make up his hand. The rest is history. To this day, x-ray is a vital tool in radiography that allows us to visualize bone and internal tissues. It was made possible only because Professor Roentgen took the translational step and applied his discovery to his own body.[4]

4.2.3 Promotion of Multidisciplinary Collaboration

Translational science is a challenging endeavor because it requires the expertise of diverse disciplines. Drug discovery alone requires expertise from physicians, biochemists, pharmacists, chemical engineers, regulatory experts, project managers, and business development – to name a few. Though the multidisciplinary nature of the work greatly increases complexity, the merits that come with the pursuit are immeasurable, as collaboration among experts yields discoveries and various applications in medicine that would not occur otherwise. There are numerous examples. First, we have optogenetics, a technique that has led to countless discoveries on cellular mechanisms since the late 1970s. Optogenetics involves controlling the activity of cells in vivo through light, manipulating light-sensitive ion channels on the surface of target cells. Optogenetics contributed to the discovery of mechanisms behind cardiac tissues, stem cells, and various regions of our brains.[5] However, the discovery of this technique and the long list of discoveries that followed would not have been possible without the experience gained from lasers, optics, photomedicine, biology, and biochemistry, each of which contributed unique insight that facilitated this development. This process works because one expert covers another's gap in knowledge and vice versa, overcoming flaws in methodology and greatly expediting the speed of development. If an electrical engineer with no background in neuroscience and biochemistry were asked to develop the method behind optogenetics and conduct experiments to yield discoveries of neural circuits in the amygdala, it would take them significantly longer to pursue this endeavor alone than if they were to team up with a neuroscientist and a biochemist. The additional experts can give their insight built from years of consolidating knowledge in their own respective fields, allowing the engineer to rely on their judgement rather than to learn from the ground up. Translational science is able to bring together many experts from different disciplines under one objective to improve patient outcomes with new drugs or devices. This unique collaboration is what makes it so valuable in the scientific community.

The second most important benefit of translational science is that it provides unmatched education for all parties. It provides a collaborative platform where scientists from diverse disciplines can congregate, teach others about their field, and bounce ideas off of one another. This unique educational environment provides a gateway to more interdisciplinary projects, ultimately leading to a positive feedback cycle of initiating more translational studies. Translational science also builds and reinstates a culture of collaboration, which makes it easier to initiate interdisciplinary work with other diverse disciplines.

4.3 CHALLENGES OF TRANSLATIONAL SCIENCE

Without a doubt, many challenges are associated with translational science. It is important to keep these challenges in mind as we try to make translational science more commonplace and accessible.

4.3.1 Complex Processes and Extended Length of Time

Translational science is crucial because it prompts scientists to conceptualize clinically relevant applications and realize their potential. It prompts scientists to take the next step in their basic science observations and apply their findings to medicine. Granted, if given enough time, another scientist could carry on the task to make clinically relevant applications. However, waiting for another scientist to take on the role leaves a chance that the translation is never realized or is significantly delayed. This is due to the fact that the scientific phenomenon could lose traction in the scientific community over time. Also, translating a discovery is a complicated and drawn-out endeavor, which can discourage many from taking on the responsibility in the first place.

The translational process of the external cardiac defibrillator is a great example of this. In the early 1900s, General Electric switched its electrical transmission from direct current to alternating current

(AC) and noticed that more of its linemen were getting killed onsite. To understand this phenomenon, General Electric funded a research study led by William Kouwenhoven and Guy Knickerbocker at Johns Hopkins University. In 1933, Kouwenhoven and Knickerbocker discovered that when stray dogs were struck with AC current, they were electrocuted, which led to their death. However, the researchers also discovered that administering a second AC current, known as a countercurrent, brought the dogs back to life. In a perfect universe, Kouwenhoven and Knickerbocker would test this phenomenon on other animals, create a prototype for a cardiac defibrillator, and successfully demonstrate it on actual patients within a few years of their discovery. However, this was not the case. It was not until 1950, 17 years after their discovery, that the first prototype of a cardiac defibrillator was created, and not until 1957, another seven years, that the first successful human demonstration occurred. Even the first successful case would not have taken place if it was not for Gottlieb Friesinger, a resident physician and lab assistant to Kouwenhoven, who had tested the prototype on a dying patient without proper authorization. After a patient who came into the emergency room collapsed on the floor, Friesinger ran up to Kouwenhoven's lab, convinced a security guard to let him inside with the prototype defibrillator, and administered two shocks, restarting the patient's heart. Who knows how much longer it would have taken for the first human trial if it was not for Friesinger's determination.[6]

4.3.2 Immature Technology and Techniques Inhibiting Translational Work

Another important obstacle in translational science is the lack of translational tools and technology needed to implement findings in human clinical trials. These tools are often necessary, and without them, translational work can be dramatically slowed down.

Professor Roentgen was fortunate to translate his observation of x-ray into a clinically relevant discovery. However, many scientists who discovered natural phenomena were not the ones to translate them to clinically relevant applications. One such case is that of Scottish physician and microbiologist Sir Alexander Fleming. After returning from vacation in 1928, Dr. Fleming went into his lab to find his Petri dishes of *Staphylococcus aureus*, a bacterial species known to cause abscesses and infection, contaminated with mold. He noticed that the area around the mold in the Petri dish was clear of staph bacteria. Upon inspection under the microscope, he discovered that the mold had inhibited the normal growth of new *S. aureus* colonies. We now know why this is the case: penicillin inside the mold inhibits gram-positive bacterial cell wall synthesis. However, back then it was not clear why this happened. Intrigued by this phenomenon, Dr. Fleming took a few more weeks to grow enough mold to confirm his findings. He was able to publish his work but was not able to translate his discovery into a working antibiotic. There was no method at the time to isolate or purify penicillin for safe antibiotic injection.[7]

The translational work to adapt Dr. Fleming's observations was carried out by Drs. Florey and Chain, together with their colleagues at Oxford, in 1939, a decade after the discovery of penicillin. Dr. Florey's team was on a mission to isolate penicillin from the mold for clinical trials. They used their laboratory as a space to grow and ferment around 500 liters of penicillin mold. Biochemist Norman Heatley filtered the penicillin by extracting the filtrate into amyl acetate and water. But his effort was not enough to carry out a large-scale clinical trial. Edward Abraham then used alumina column chromatography, a newly developed isolation technique, to purify penicillin and ramp up production. The following year, Dr. Florey's team conducted a clinical trial that showed the role of penicillin in protecting against streptococci bacterial species in mice. The first human clinical trial was conducted a year later when Albert Alexander, a policeman with a life-threatening infection, became the first recipient of penicillin. Following the injection, Albert Alexander showed signs of recovery, but the stockpile depleted, resulting in his death.

Dr. Florey's team is credited for translating the work of Dr. Fleming. Although Dr. Fleming had noted the potential therapeutic implications of penicillin in his paper, he was not the one to translate his own findings. Again, the technology and techniques at the time did not exist; therefore, translation was impossible. Today, we are faced with a similar conundrum. Over 90% of investigational drugs that enter human clinical trials do not make it to market, and one of the biggest obstacles facing scientists is the insufficient tools and methods to accurately translate science to medicine. Animal studies and in vitro modeling methods such as microfluidic organ-on-chip exist to offset this difficulty, but they are not enough to accurately predict human biochemical responses to drugs. For an effective translational study to take place more easily, more effort is needed in developing these tools that underlie drug development, not just the translational effort itself.

4.3.3 Challenges in Teamwork and Management

As discussed in Section 4.2, the translational endeavor is a team effort that requires multidisciplinary collaboration. If such collaboration is effective, the study can yield great success. However, we must understand that more hands-on deck means more potential points of failure.

Most musicians will agree that a conductor is crucial to the functioning of an orchestra. Without a conductor, the tempo of a piece will be asynchronous and each individual in the orchestra can dictate on their own accord. Here, the players of the instruments are the individual key members of the translational science endeavor and the conductor is the principal investigator or a project manager, overseeing the whole operation. At one stage of translational science, there may be several pursuits happening simultaneously. If this is a drug discovery process, chemical engineers may be finalizing the synthesis of an investigational drug for Phase 1 clinical trials. Meanwhile, laboratory scientists may be finalizing their animal studies and analyzing their findings. Concurrently, the team may be actively recruiting patients for Phase 2 clinical trials while regulatory team members obtain IRB or FDA approval. It is up to the principal investigator to effectively manage the individual pursuits to avoid chaos. If the project management is not effective, significant delays may ensue in each step, extending the deadline even more. For instance, if the team is applying for IRB approval to conduct animal or human trials, it must meet the submission deadline set by the institution. The submission deadline usually happens once or twice a month. If there are issues in teamwork and the team fails to meet the deadline, the team will have to wait until the next available submission date. To avoid issues and miscommunication, each individual must have a clearly defined role and the initiative to fulfill their responsibilities. They must be able to communicate appropriately and frequently should any problems arise. This seems simple in writing, but effective teamwork is often difficult to achieve.

Another aspect of teamwork unique to translational research is the level of collaboration that is required from different experts. It is important to establish the relationship and consult these experts to lay out a detailed plan early on. One of the core members critical to this team is a physician in the field of practice pertinent to the research. Physicians have an in-depth level of understanding of the medical landscape, including the types of therapies that exist in the market, standard of care, and patient needs. They are also familiar with the disease pathophysiology and can address clinically relevant foundations needed in obtaining grants. A physician can help guide the research and identify the target patient population. Furthermore, they can consult during the early design stage to advise on which therapies or devices can and cannot work. Last but not least, it is also important to have a physician sponsor for the clinical trials. Physicians well connected to an academic hospital network have access to patients for recruitment and can lead the clinical studies. However, it can be difficult to find a physician collaborator who has dedicated time to research.

4.3.4 The Importance of Advanced Training

By far, the most difficult challenge of translational science is the years of training required for properly leading translational research. The principal investigator(s) must be well versed in many domains and skill sets. This includes being a domain expert in their field of study – not only mastering the fine details but also recognizing the bigger picture and developments in their field. This is not much different from a basic scientist who only partakes in lab bench work. To conduct bench to bedside research, however, translational researchers must go the extra mile. They need to be familiar with the current practice of medicine, at least in the particular field or specialty into which they are trying to introduce the translational work. For instance, if a translational scientist is developing a new lung cancer drug based on a treatment mechanism they had observed in the lab, they should be familiar with the existing cancer drugs or antineoplastics approved for use. How is their new drug different from the existing drugs? What are the anticipated side effects of stimulating or inhibiting certain pathways? They should also be familiar with the pathophysiology behind neoplasms or cancer development. What is the mechanism of neoplastic progression and metastasis? What grade and stage of lung cancer is their therapy targeting?

Last but not least, it is essential that translational researchers are familiar with regulatory affairs. This entails having a deep understanding of the FDA drug approval process and clinical trials. Having this understanding is crucial for lengthy processes like drug discovery, as successful drug approval depends heavily on foresight and planning.

Another fundamental skill of a translational scientist is research. This involves asking clinically relevant questions, forming a strong hypothesis, devising accurate methods, and analyzing data to maximize reproducibility and credible science. To develop these skills, translational scientists need extensive research training and experience conducting and leading independent research studies. Obtaining domain-specific knowledge, medical knowledge, and research skills requires a significant investment in time in training alone. For one to obtain this level of mastery, a decade of post-secondary schooling and research training will be needed. Translational researchers must also master the different types of research that happen under translational work. First is the basic science or engineering research, including drug or device candidate identification and optimization. Second is the clinical research component, including trial design, regulatory compliance, and safety and efficacy analyses.

4.3.5 The Reality of Translational Research in the Private Sector and Academia

Translational research can be carried out in the private or the public sector. These two sectors do collaborate, but for now, we will keep them separate. In the private sector, biopharmaceutical companies run translational research in the hopes of yielding a new drug, therapy, or device that can increase their revenue and profits. The public sector comprises academic institutions and other non-profit institutes that are mostly funded by tax revenues and donations. The main difference between the two sectors is the nature of translational research. Major pharmaceutical companies are obligated to work for the shareholders, who are driven by profit. This means that the company executives are very selective with their pursuits because funding a single project can last for decades and cost billions of dollars. The pursuit that they choose to invest in therefore needs to yield a greater return than their initial investment. This means two things. First, the private sector is reluctant to fund research studies that have limited upside. Rare diseases with low prevalence or diseases that have limited demand for therapy may be excluded. The Orphan Drug Act of 1983 tried to offset this balance by facilitating the development of drugs to treat rare diseases, but the natural instincts of the private sector gear them toward diseases with high prevalence and demand. Second, the private sector is reluctant to fund projects that are deemed too risky. Every drug discovery is risky in one way or another, and so the risk is evaluated by calculating whether there is enough data to back the pursuit. This means inspecting published data pertinent to the therapy target and mechanism and consulting experts in the field to determine the odds of success. This can mean limited freedom to pursue a disease of interest for investigators working in the private sector.[8]

However, this does not mean that the private sector is not fit to conduct translational research. The private sector can be an incredibly efficient vessel for drug discovery once a target is chosen. Johnson & Johnson, for example, exceeds the market cap of $400 billion and can act as its own nation state with over 130,000 employees. In 2020, Johnson & Johnson spent over $12 billion on research and development alone.[9] One overwhelming advantage of company-led translational research is that resources and efforts can be *concentrated* in the drug discovery process, optimizing time and resources every step along the way. We will soon see that the public sector is not readily equipped to do so. This may be the reason that the private sector triumphed in the race to develop COVID-19 vaccines in such a short timeframe. The private sector is also able to outpace the public sector in manufacturing and disseminating newly discovered therapies, which pertains to the objectives of the T3 stage of translational science. Companies can create a better distribution network by analyzing and building cost-saving logistics, ultimately reaching more patients.

The public sector, on the other hand, may have greater freedom as it does not respond to financial incentives and is mostly funded by the federal government. The public sector can fund research involving rare diseases that may not necessarily attract the private sector. With respect to the goal of translational research – producing applicable interventions that enhance patient outcomes – the public sector embodies it more fittingly. However, this does not mean that the public sector is flawless. Research in the public sector is mostly run by two entities. First, academic universities hire independent researchers who obtain funding on their own. Second, institutes also hire independent researchers but are concentrated on a few objectives or diseases. For instance, the National Institute of Neurological Disorders and Stroke primarily focuses its efforts on spinal cord injuries and repair and will hire investigators who align with its purpose.[10] In academic universities, a high level of independence is granted to researchers with respect to the type of research conducted, but independence comes with the downside of strict adherence to meritocracy. Academic research professors and investigators are pressured to showcase productivity. That productivity can be shown through the number of publications published in high-impact journals, the number of citations, and the amount of funding and grants brought to the respective universities. The phrase "publish or perish" can be a looming reality in academic research.

Translational research is a difficult endeavor that requires dedication from multiple researchers spanning many disciplines and several years. Unfortunately, the success of yielding a new drug or therapy is also low. The success metric of academic research that strictly relies on publications, citations, and funding can deter many early-career scientists from embarking on the journey of translational research even though the potential outcome may be rewarding. Even physician scientists who are well equipped to conduct translational research are put under the same pressure as their PhD colleagues. Dr. Mark J. Eisenberg puts this well in his book *The Physician Scientist's Career Guide*.[11] "Clinicians are typically given 3–5 years to get a grant and show research productivity through publications. If the academic hospitals are not satisfied, they will require you to change your role as a clinician-educator, where you primarily care for patients and teach medical students". Therefore, at the end of the day, translational research is limited to the handful of established researchers who not only have received the necessary advanced training to conduct translational research but are also certain that their translational research can surpass the productivity metric set by their institution.

4.3.6 Challenges Associated with Engineering-Related Translational Research

One challenge uniquely associated with translating engineering discoveries to medicine lies in the fundamental differences between the two disciplines. When approaching a new problem, engineers tend to think about what is possible. The natural instinct of an engineer is to disrupt the status quo and find better ways to solve a problem. Physicians, on the other hand, look at what has already been done – they use the tried-and-true method of reviewing a stockpile of clinical data. Physicians are the sentinels of patient safety and only allow innovation when sufficient evidence is presented. These two lines of thought can clash when engineers overextend on a solution or when physicians become overprotective and disbar novelty. And the process can quickly be put in a deadlock. Successful translational research can only happen when there is a fine balance between the two axioms. It is important that engineers respect the clinical expertise and acumen of physicians and shape their solutions to ensure patient safety. On the other hand, it is imperative that physicians let engineers innovate when they see the opportunity for it.

In the design stage of the SSA, the engineers had some difficulty understanding the neurosurgeons' perspective and their clinical acumen due to the engineers' lack of surgical knowledge and experience. Due to this disparity in knowledge, it presented a challenge for neurosurgeons to explain why certain solutions do not work in the operating room. Some aspects of the operating room that seemed very clear and intuitive to the surgeons were not so obvious to the engineers. For instance, the engineers envisioned a feature in which the SSA could connect to Google Scholar or PubMed to search for answers to questions that surface during a surgical procedure. The engineers based this idea on the countless number of times they used resources on the internet when fixing a circuit or debugging a code and thought that this feature was a practical addition that could bring in valuable outside resources to the operating room. What the engineers did not realize is that the surgeons do not elicit the help of the internet when they have a question. Even before the operation commences, the operating surgeon has a detailed surgical plan and procedural movements mapped out on their head. They also usually have several contingency plans laid out. Therefore, it is very rare that they need the help of outside resources intraoperatively.

The challenges of translational science are unique and can be difficult to overcome. However, these challenges should not discourage early-career scientists from pursuing transitional science. Many resources, including this one, may serve as a guide to navigating the difficult process of translational science. Although it is difficult, translational science is a rewarding endeavor. Once it is successfully "translated", the work will hopefully have a long-lasting impact on patient care.

REFERENCES

1. Waldman, Scott A., and Andre Terzic. "Clinical and translational science: From bench-bedside to global village." *Clinical and Translational Science* 3.5 (2010): 254.

2. DiMasi, Joseph A., Henry G. Grabowski, and Ronald W. Hansen. "Innovation in the pharmaceutical industry: New estimates of R&D costs." *Journal of Health Economics* 47 (2016): 20–33.
3. Greatbatch, Wilson. *The making of the pacemaker: Celebrating a lifesaving invention.* Prometheus Books, 2011.
4. Frankel, Richard I. "Centennial of Röntgen's discovery of x-rays." *Western Journal of Medicine* 164.6 (1996): 497.
5. Deisseroth, Karl. "Optogenetics." *Nature Methods* 8.1 (2011): 26–29.
6. Boveda, Serge, Stéphane Garrigue, and Philippe Ritter. "The history of cardiac pacemakers and defibrillators." *Dawn and evolution of cardiac procedures.* Springer, 2013, 253–264.
7. Bennett, Joan W., and King-Thom Chung. "Alexander Fleming and the discovery of penicillin." *Advances in Applied Microbiology* 49 (2001): 163–184.
8. Austin, Christopher P. "Opportunities and challenges in translational science." *Clinical and Translational Science* 14.5 (2021): 1629–1647.
9. Johnson & Johnson. *2020 annual report.* March, 2021.
10. Roberts, J., S. Waddy, and P. Kaufmann. "Recruitment and retention monitoring: Facilitating the mission of the National Institute of Neurological Disorders and Stroke (NINDS)." *Journal of Vascular and Interventional Neurology* 5.1.5 (2012): 14.
11. Eisenberg, Mark J. *The physician scientist's career guide.* Springer Science & Business Media, 2010.

OVERCOMING THE CHALLENGES OF TRANSLATIONAL RESEARCH

In this chapter, we are going to address the challenges associated with translational science laid out in the previous chapter. It is important to remember that the best approaches to address these challenges are heavily debated, and there is no one correct way.

5.1 EXPANSION OF TRANSLATIONAL SCIENCE EDUCATION AND RESEARCH

One of the first challenges that we addressed in the previous chapter is associated with the complexity and the time it takes to conduct translational research. One way to address this challenge is to think about whether the process could be simplified in any way – for example, allocating more resources and funding to optimize the translational science process. Do not confuse this with simply increasing funding. We are emphasizing the need to research the discipline and the act of translating itself. How do we best improve the efficiency and turnover between the T1 and T2 stages of the drug discovery process? How do the translational research processes of drug discovery and medical devices differ, and what adjustments need to be made to address these differences? These are the types of questions to answer to facilitate translational research and improve the efficiency of the whole process.

The National Center for Advancing Translational Sciences (NCATS) was formed in 2010 and has been instrumental in addressing translational science endeavors. The mission of NCATS is "to catalyze the generation of innovative methods and technologies that will enhance the development, testing and implementation of diagnostics and therapeutics across a wide range of human diseases and conditions". NCATS has worked hand-in-hand with the Clinical and Translational Science Awards given by the National Institutes of

DOI: 10.1201/9781003253341-5

Health. One key initiative supported by NCATS relates to educational opportunities for translational research. NCATS offers various training opportunities within the agency and in partner institutions that focus on nurturing translational scientists in varying stages of their careers. From summer internship programs to postdoctoral fellowships, NCATS offers unique focused training opportunities. NCATS also provides courses focused on setting the foundation for preclinical translational science. The initiatives taken by NCATS address and amplify the needs of the translational science community.[1]

However, more could be done. Each seven-week online course offered by NCATS costs roughly $350. For students and trainees, this may be a hefty price tag. NCATS could follow the example set by Massachusetts Institute of Technology and make these courses freely available to students and trainees. This small bid could exponentially increase the number of people watching the courses, which may disseminate the knowledge more quickly. Another way to enhance the training courses would be to include a seminar or lecture series that hosts translational researchers from diverse fields. It would be a unique platform for translational scientists to publicize their research and deliver their insight and expertise in translational science. The speakers could address the difficulties they encountered in their translational research and describe how they overcame them. These would be invaluable lessons for the scientific community, as listeners could learn from the speakers' mistakes and successes.[2]

Another improvement would be to create a mentorship program at a national level for researchers at different stages. A researcher who signs up for the program can elect to mentor less experienced scientists or receive mentorship from more experienced scientists. This could be a great way for translational scientists to receive long-term guidance and form new and meaningful connections with scientists across the spectrum.

5.2 INCREASED FUNDING IN TRANSLATIONAL RESEARCH APPARATUS AND METHODS

If we are given a task to scavenge for dinosaur fossils in the middle of a desert, we would be better equipped with a forklift than a spoon. Likewise, drug discovery and medical device research can greatly benefit from tools that significantly expedite the process. There are many promising tools in development today that can make it happen. First, there are tissue-on-chip and organ-on-chip platforms that facilitate drug discovery by mimicking the cell–tissue interaction in vitro. Scientists are able to use these platforms to accurately model the drug pharmacological effects on an affordable platform. Liver-on-chip technology can also screen for drug toxicity to determine the safety of an experimental drug.[3] This can not only save research expenditure but can also save time by reducing the chance of failure in human clinical trials. Next are the developments in imaging technology that can visualize our bodies faster and more accurately than before. One of these is near-infrared laser speckle imaging, which can precisely image blood vessels and capillaries. These imaging modalities can accurately depict real-time effects of drugs and medical devices.[4] Last but not least, leaps in AI have large implications for drug discovery. Isomorphic, a company started under Alphabet, is developing an AI algorithm based on DeepMind's protein-folding predicting ability to scan through molecules that can suitably act on a target. This is a significant improvement over the trial and error method used to identify fitting molecular shapes and saves time in drug candidate isolation. AI is also put to work to predict the drug-to-drug and drug-to-enzyme interaction, recognizing drug toxicity in advance. Ultimately, funding these research projects that accelerate translational processes provides a great return, as more translational research can be conducted in a shorter time frame.

5.3 CONTINUED INVESTMENT INTO PHYSICIAN SCIENTISTS

In the previous section, we mentioned that advanced training is required to pursue translational sciences. Specifically, we noted that translational scientists need domain-specific knowledge, research-specific medical knowledge, and extensive research experience. Although there are ways to shorten the process of translational research by building innovative tools discussed earlier, there are unfortunately no shortcuts to education and training. That being said, there are training programs in the US that can produce exceptional translational scientists because they provide the necessary training and perspective one needs to conduct translational research. Those programs are the MD/PhD degree programs.

The MD/PhD program is a combined degree program designed to nurture physician scientists or physician engineers. A subset of MD/PhD programs is referred to as the Medical Scientist Training Program, which denotes sponsorship by the National Institute of

General Medical Sciences of the National Institutes of Health. Most of the program is split into the 2–4–2 curriculum, in which the first two years are the preclinical training done in medical school, followed by four years of the PhD curriculum, and concluding with the clinical training typically done in the last two years of medical school. This program is vital to nurture translational scientists because it gives one the ability to conduct all stages of translational research. First, the physician scientist has domain-specific knowledge and the research skills to come up with relevant pursuits in their field. They can also define the treatment mechanisms and targets for drug candidate or device development. Having had dedicated research time to pursue an independent project, physician scientists are well equipped to plan and execute good research. Once physician scientists become attending physicians, they can treat patients and collect valuable data for research. They also can easily launch clinical studies with access to patients. Thanks to their medical training, they are well versed in the medical landscape and know the treatments or procedures, standard of care, and insurance mechanism to guide the dissemination process of their translational research. In 2016, there were 602 MD/PhD graduates in the US. However, this is half the number of people needed to meet the projected demand of physician scientists.[5] With the growing importance of translational science, more funding and training grants should be allocated to training physician scientists.

It is easy to ask for more funding for translational research. However, it is more important that we devise specific plans of action that can greatly expand the current efforts. Although these methods are well grounded, they are not comprehensive. We call on our readers to ponder on new ideas and methods to spread the importance of translational research. As this book explored in Chapter 4, translational science is the most important avenue for healthcare revolution, and our effort to create the smart surgical assistant should not be the only endeavor to bring innovation and change to medicine. We end this book hoping for a future that adopts new ways to enhance translational research and a scientific community that is deeply invested in it.

REFERENCES

1. National Institutes of Health. "National center for advancing translational sciences." http://www.ncats,nih.gov.

2. "Translational science training and educational resources." National Center for Advancing Translational Sciences, U.S. Department of Health and Human Services, https://ncats.nih.gov /training-education/resources.

3. Khetani, Salman R., and Sangeeta N. Bhatia. "Microscale culture of human liver cells for drug development." *Nature Biotechnology* 26.1 (2008): 120–126.

4. Dunn, Andrew K. "Laser speckle contrast imaging of cerebral blood flow." *Annals of Biomedical Engineering* 40.2 (2012): 367–377.

5. Kosik, R. O., et al. "Physician scientist training in the United States: A survey of the current literature." *Evaluation & the Health Professions* 39.1 (2016): 3–20.

INDEX

Printed in the United States
by Baker & Taylor Publisher Services